Springer Theses

Recognizing Outstanding Ph.D. Research

Aims and Scope

The series "Springer Theses" brings together a selection of the very best Ph.D. theses from around the world and across the physical sciences. Nominated and endorsed by two recognized specialists, each published volume has been selected for its scientific excellence and the high impact of its contents for the pertinent field of research. For greater accessibility to non-specialists, the published versions include an extended introduction, as well as a foreword by the student's supervisor explaining the special relevance of the work for the field. As a whole, the series will provide a valuable resource both for newcomers to the research fields described, and for other scientists seeking detailed background information on special questions. Finally, it provides an accredited documentation of the valuable contributions made by today's younger generation of scientists.

Theses are accepted into the series by invited nomination only and must fulfill all of the following criteria

- They must be written in good English.
- The topic should fall within the confines of Chemistry, Physics, Earth Sciences, Engineering and related interdisciplinary fields such as Materials, Nanoscience, Chemical Engineering, Complex Systems and Biophysics.
- The work reported in the thesis must represent a significant scientific advance.
- If the thesis includes previously published material, permission to reproduce this must be gained from the respective copyright holder.
- They must have been examined and passed during the 12 months prior to nomination.
- Each thesis should include a foreword by the supervisor outlining the significance of its content.
- The theses should have a clearly defined structure including an introduction accessible to scientists not expert in that particular field.

More information about this series at http://www.springer.com/series/8790

Kirsty Elizabeth Duffy

First Measurement of Neutrino and Antineutrino Oscillation at T2K

Doctoral Thesis accepted by
University of Oxford, UK

Springer

Author
Dr. Kirsty Elizabeth Duffy
Mail Stop 220,Fermilab
Batavia, Illinois
USA

Supervisor
Prof. Alfons Weber
Pembroke College
University of Oxford
Oxford
UK

and

Science and Technology Facilities Council
Rutherford Appleton Laboratory
Oxford
UK

ISSN 2190-5053 ISSN 2190-5061 (electronic)
Springer Theses
ISBN 978-3-319-65039-5 ISBN 978-3-319-65040-1 (eBook)
DOI 10.1007/978-3-319-65040-1

Library of Congress Control Number: 2017950006

Printed on acid-free paper

This Springer imprint is published by Springer Nature
The registered company is Springer International Publishing AG
The registered company address is: Gewerbestrasse 11, 6330 Cham, Switzerland

To my parents

Supervisor's Foreword

"Sehr geehrte radioaktive Damen und Herren"[1],

This is the opening of the letter in which the Austrian physicist Wolfgang Pauli gave birth to the neutrino. At the time, he postulated a neutral particle, the neutrino, to solve a crisis in nuclear physics. It was observed that in certain types of radioactive decays neither energy nor angular momentum seemed to be conserved. To solve this problem, he invented an "invisible" particle that would carry some energy and angular momentum away, thus solving one problem by introducing an undetectable solution. Is that called progress?

In the end, Pauli was wrong in his assumption that the neutrino could not be measured; it was discovered almost 30 years later by the famous Reines and Cowan Experiment close to a nuclear reactor. Pauli was very right: his "invisible" particle is needed and does exist.

Already in its initial inception, it became clear how closely the neutrino is related to our understanding of the very basic and fundamental laws of physics. Many other instances followed where the neutrino helped to understand the very fabric of our universe: parity violation in the weak interaction, studying nuclear fusion in the sun and discovering that the "massless" neutrino actually does have mass and can change flavour from one type to another. First hints have evolved that the symmetry between matter and antimatter might be violated in neutrino oscillations and that they may help us to understand why there is more matter than antimatter in the universe. These are just a few of the interesting aspects of the neutrino that touch on the fundamental laws of our existence. And it is this last point, neutrino and antineutrino oscillations and the search for CP-violation[2] in neutrino physics, which is the topic of this thesis.

Over the last 20 years, there has been mounting evidence there is something wrong with the neutrino, which has been thought of as massless, neutral, spin-$\frac{1}{2}$

[1]German for "Dear radioactive ladies and gentlemen"

[2]In this context, CP-violation is the difference between the oscillation properties of neutrinos and antineutrinos, a matter-antimatter difference.

particle that can only interact via the weak force (and gravity). Three distinctive neutrino species have been discovered, named electron, muon and tau neutrinos after the charged lepton they are related to. Studying the flux of neutrinos coming from the nuclear fusion processes in our sun or the neutrino flux produced in the upper atmosphere by the bombardment of cosmic rays turned up a few surprises. Only $\frac{1}{3}$ of the electron neutrinos produced in the sun reach us here on earth, and roughly $\frac{1}{2}$ of the muon neutrinos produced by cosmic rays have been detected. The Nobel Prize in physics 2015 was awarded jointly to Art McDonald and Takaaki Kajita for proving that neutrinos can change from one flavour to another, which is only possible if neutrinos have mass.

It is now firmly established that neutrinos not only have mass, but also that the flavour transition is due to neutrino oscillations, a process in which neutrinos go from one type to another and back. This is a quantum mechanical process caused by the flavour and mass eigenstates not being identical. These states are related by a unitary mixing matrix called the PMNS matrix. Over the last decade, the parameters governing this process have been measured with increasing precision, not only for neutrinos, but also for antineutrinos. Differences between neutrino and antineutrino oscillations can be caused by a complex phase in the PMNS matrix. The measurement of this complex phase is currently the holy grail of neutrino physics and one of the key measurements reported in this thesis.

Dr. Duffy gives a very educational introduction into the basics of neutrino physics with an emphasis on long-baseline neutrino oscillations and the T2K experiment, which puts the oscillation measurement and the search for CP-violation into context.

Understanding the uncertainty is an essential part of making any measurement. Often, systematic effects are added in the analysis as constrained parameters in a high-dimensional fit that model the uncertainties of the measurement. Finding the minimum of the fit and their uncertainties, and therefore the physics result, is highly non-trivial. This thesis introduces the technique of Bayesian Interference and the Markov Chain Monte Carlo method to solve this problem. In particular, this more technical aspect, which can be applied to quite different problems in physics and beyond, makes this thesis an educational read.

Those interested in the physics of neutrino oscillations will learn how a single experiment has constrained the CP-violating phase δ_{CP} and excluded CP-conservation at 90% probability, the first step in a large worldwide endeavour.

This thesis can act as reference for beginners and experts in neutrino physics alike. It lays out the background to an exciting area of particle physics and introduces the physics and techniques that can be used to analyse the data.

Oxford, UK
June 2017

Prof. Alfons Weber

Abstract

The T2K experiment is a long-baseline neutrino oscillation experiment, in which a muon neutrino beam is directed over a 295-km baseline from the J-PARC facility to the Super-Kamiokande detector. This allows neutrino oscillation to be studied in two channels: disappearance of ν_μ and appearance of ν_e. The T2K beam can be run in either neutrino or antineutrino configuration, for a neutrino beam which is predominantly composed of ν_μ or $\bar{\nu}_\mu$ respectively.

This thesis presents the first oscillation analysis to combine neutrino-mode and antineutrino-mode data sets, giving the first ever sensitivity to the CP-violating phase δ_{CP} from T2K data alone, as well as the most precise T2K measurement of the other neutrino oscillation parameters. The analysis uses a Markov Chain Monte Carlo method to construct a sample from the Bayesian posterior distribution, from which the oscillation parameters of interest and their uncertainties are estimated. Data samples from the T2K near detector, ND280, are fit simultaneously with data from the far detector in order to reduce the uncertainty in the far-detector prediction.

When fitting the T2K data alone, the best-fit oscillation parameter values are $\sin^2\theta_{23} = 0.519^{+0.031}_{-0.059}$, $\sin^2\theta_{13} = 0.0257^{+0.0106}_{-0.0440}$, $\Delta m^2_{32} = 2.54^{+0.12}_{-0.10} \times 10^{-3}\,eV^2$ and δ_{CP} = -1.91 rad. The 90% credible interval for δ_{CP} excludes values around $\pi/2$: $\delta_{CP} \notin [0.38, 2.60]$ rad. When fitting the T2K data with a constraint on $\sin^2 2\theta_{13}$ from measurements by reactor experiments, the best-fit oscillation parameter values are $\sin^2\theta_{23} = 0.519^{+0.061}_{-0.029}$, $\sin^2\theta_{13} = 0.0219^{+0.0014}_{-0.0014}$, $\Delta m^2_{32} = 2.54^{+0.12}_{-0.12} \times 10^{-3}\,eV^2$ and δ_{CP} =-1.72 rad. The 90% credible interval for δ_{CP} contains values $\delta_{CP} \in [-3.10, -0.17]$ rad, excluding the CP-conserving values 0 and $\pm\pi$ at 90% probability.

Acknowledgements

The work presented in this thesis would not have been possible without the help and support of many people. Of course, I would like to start by thanking my supervisor, Alfons Weber, for his support and advice throughout my D.Phil. and for giving me the many opportunities that have led to this point. Thanks also to the rest of the Oxford neutrino group: to Giles and Dave for their additional support, both practical and intellectual, and to my fellow students—Stephen, Raj, David, Tom, Zac, Abe, Tom and Charles—for their friendship and for making the office a fun place to work. The coffee breaks, lunches, pub quizzes, karaoke and visits to Shinchan's have made the last four years a very enjoyable experience.

This thesis has been through many revisions since its first drafts. I would like to thank Alfons Weber and Asher Kaboth for their helpful comments, and my D.Phil. examiners, Armin Reichold and Peter Dornan, for their enlightening discussions and useful suggestions.

Two aspects of my work during my D.Phil. have unfortunately not made it into this thesis—developing a time calibration for the ECal and serving as a DAQ expert. For their help and support with that work, many thanks are owed to the DAQ group (particularly Helen, who is always a Skype call away) and calibration group (especially the conveners, Jen Haigh, Yevgeniy Petrov, Per Jonsson and Istvan Danko).

The result of an oscillation analysis is rightly viewed as an achievement of the entire collaboration, and I owe a lot to the T2K collaboration for their input and advice. I would especially like to thank the oscillation analysis group, and in particular the conveners of the group while I was working in it, Atsuko Ichikawa, Mark Hartz and Claudio Giganti, for their invaluable help and guidance.

This analysis was developed as part of the MaCh3 team on T2K, and I am grateful to everyone that has worked on the software or worked with me on the analysis. Special thanks are owed to Asher Kaboth, for his friendship and support. Throughout my analysis work, Asher has been constantly available to offer advice on any topic, from the minutiae of the code to the big-picture interpretations of results, and I couldn't have done this without him.

I would like to acknowledge that the work presented here made use of Emerald, a GPU-accelerated high-performance computer, made available by the Science & Engineering South Consortium operated in partnership with the STFC Rutherford-Appleton Laboratory.

Finally, I would like to thank all my family and friends, for their continued support throughout the D.Phil. In particular, thank you to my parents, Luke, and anyone else that helped me stay sane during the dreaded writing-up process—there are far too many people to name, but I hope you know who you are.

Contents

1 Preamble 1
References. 3

2 Introduction to Neutrino Oscillation 5
2.1 The Discovery of the Neutrino 6
2.2 Neutrino Oscillation 8
2.2.1 Initial Evidence for Neutrino Oscillation. 9
2.2.2 The Theory of Neutrino Oscillation 11
2.2.3 Current World Knowledge and Open Questions 21
References. 26

3 The T2K Experiment 31
3.1 The T2K Beam. 32
3.1.1 The T2K Neutrino Beamline. 33
3.1.2 The Off-Axis 'Trick'. 35
3.1.3 Neutrino Flux Simulation 36
3.2 The T2K Near Detectors 37
3.2.1 INGRID 38
3.2.2 ND280 39
3.2.3 Near Detector Simulation 45
3.3 The T2K Far Detector: Super-Kamiokande. 45
3.3.1 Super-Kamiokande Simulation 48
3.4 Neutrino Interactions at T2K 48
3.4.1 Neutrino Interaction Simulation. 52
3.5 Data Analysed 53
References. 54

4 Bayesian Inference and the Markov Chain Monte Carlo Method. 57
4.1 Bayesian Inference 58
4.2 Monte Carlo Methods. 59

4.3 Markov Chain Monte Carlo 60
4.3.1 The Metropolis-Hastings Algorithm 61
4.3.2 Proposal Function and Step Size. 62
4.3.3 Burn-In 65
4.4 Extracting Information from the Posterior. 66
4.4.1 Marginalisation of Nuisance Parameters 66
4.4.2 Point Estimation 69
4.4.3 Uncertainty on Point Estimates 70
4.4.4 Model Comparison 73
4.4.5 Posterior Predictive Method 76
4.4.6 Goodness-Of-Fit 77
References. 79

5 Joint ν+$\bar{\nu}$ Oscillation Analysis: Framework and Validations 81
5.1 Analysis Strategy 82
5.2 Event Selection. 84
5.2.1 ND280 84
5.2.2 Super-Kamiokande 87
5.3 Systematic Uncertainties. 89
5.3.1 Flux 89
5.3.2 Cross Section 90
5.3.3 ND280 Detector 94
5.3.4 Super-Kamiokande Detector 97
5.4 Monte Carlo Predictions and Pre-fit Data/MC Comparison. 98
5.4.1 Flux Tuning 98
5.4.2 Tuning for Neutrino Interaction Model 99
5.4.3 Tuning from Fits to ND280 Data 99
5.4.4 ND280 MC Prediction 99
5.4.5 Super-Kamiokande MC Prediction 101
5.4.6 Effect of Systematic Uncertainties on SK Event-Rate Prediction 104
5.5 Fitter Validation and Sensitivity 108
5.5.1 Asimov Fits 108
5.5.2 Log-Likelihood Scan. 115
References. 118

6 Joint ν+$\bar{\nu}$ Oscillation Analysis: Results. 121
6.1 Systematic Parameters 128
6.2 Comparison to Previous T2K Results. 130
6.3 Comparison to Other Neutrino Oscillation Experiments 135
6.4 Comparison to Sensitivity. 136
6.5 Effect of Prior. 139
References. 143

7 Conclusions and Outlook 145
References. 147

Appendix A: Antineutrino Oscillation: Search for $\bar{\nu}_e$ Appearance and Measurement of $\bar{\nu}_\mu$ Disappearance 149

Appendix B: Oscillation Analysis Groups on T2K 161

Curriculum Vitae 163

Glossary 169

Chapter 1
Preamble

One of the longest-running puzzles in particle physics and cosmology is the matter-antimatter asymmetry problem. Simply put: why does the universe exist? We believe that both matter and antimatter were created in equal amounts in the Big Bang, but current measurements show that matter overwhelmingly dominates the observable universe. In the Standard Model of Particle Physics, the relation between a matter particle and its corresponding antimatter particle is known as CP symmetry—if CP symmetry is violated, that means that physics draws a distinction between matter and antimatter, and could explain what causes our universe to be matter-dominated. CP violation has been observed in the mixing of quarks, but the violation observed is small and not sufficient to explain the matter-antimatter asymmetry that we see. Violation of CP symmetry is yet to be observed in the neutrino sector, but this is a promising place to look: the PMNS theory of neutrino oscillation (discussed in Chap. 2, Sect. 2.2.2) allows for large CP violation, which is parameterised in three CP-violating phases: δ_{CP}, α_1, and α_2. Only the parameter δ_{CP} can be measured in neutrino oscillation experiments such as T2K—α_1 and α_2, the so-called Majorana phases, do not affect the oscillation probability. One of the most important goals of neutrino physics in the near-to-mid future is to measure this parameter, in order to determine whether or not CP symmetry is violated in neutrino oscillation, and if it is, how strongly.

An accessible introduction to the problem of matter-antimatter asymmetry, as well as the possible role that neutrinos may have to play in this mystery, is given in Chap. 8 of [1]. Slightly more technical, but still very comprehensible, descriptions are available in [2–4], and more mathematical discussions in [5] and Chap. 11of [6].

T2K (see Chap. 3 of this thesis, or http://t2k-experiment.org) is a long-baseline neutrino oscillation experiment that was optimised to make precision measurements of the known neutrino oscillation parameters $\sin^2\theta_{23}$ and Δm^2_{32}, and to establish whether the parameter $\sin^2\theta_{13}$ is equal to zero. It was designed to run using a beam composed mostly of muon neutrinos, so measurements of neutrino oscillation can be made in two 'channels': muon neutrino (ν_μ) disappearance and electron neutrino (ν_e) appearance ($\nu_\mu \rightarrow \nu_\tau$ oscillation also occurs but τ leptons are not measured in the far detector at the neutrino energies used for T2K). After running for three years

K.E. Duffy, *First Measurement of Neutrino and Antineutrino Oscillation at T2K*,
Springer Theses, DOI 10.1007/978-3-319-65040-1_1

between 2010 and 2013, T2K published the world's most precise measurement of $\sin^2\theta_{23}$ [7], as well as an observation of ν_e appearance showing $\sin^2\theta_{13} \neq 0$ at a significance of over 7σ [8]. Following these results, focus has turned to the oscillation of antineutrinos. In 2014 the polarity of the magnetic horns used to create the T2K beam was reversed, such that π^- instead of π^+ particles are focussed in the beam direction, leading to a neutrino beam that is mostly composed of muon antineutrinos, rather than muon neutrinos.

This thesis presents an analysis of the joint effects of neutrino and antineutrino appearance and disappearance at T2K, and is the first analysis to have sensitivity to the CP-violating phase δ_{CP} with T2K data alone. The analysis described in this thesis uses data collected up to May 2016, with almost equal amounts of data taken using the neutrino-mode and antineutrino-mode beam configuration.

A glossary has been provided at the end of this thesis to assist the reader with some of the more specific terms and acronyms used in the text. Chapter 2 gives a brief overview of the history of neutrino physics, and the discovery of—and theory behind—neutrino oscillation. It also summarises the current knowledge in the field of neutrino oscillation before the analysis presented here, using information from solar, reactor, and long-baseline neutrino experiments.

The T2K experiment itself is described in detail in Chap. 3, including the beamline, detectors, and simulations used to predict the expected observables at the near and far detectors. A short introduction to neutrino interactions at T2K is also given, with information about the types of neutrino interaction observed at T2K and how they are used for the oscillation analysis.

The analysis presented here is a Bayesian oscillation analysis, which uses a Markov Chain Monte Carlo fitting technique. This is quite unusual in the field of neutrino physics: traditionally, frequentist fitting methods and interpretations have been more widely used. For this reason, Chap. 4 describes the Markov Chain Monte Carlo method, as well as the Bayesian approach to parameter estimation and interpretation of results.

The analysis itself is discussed in Chaps. 5 and 6. Chapter 5 describes the framework of the analysis: the systematic uncertainties considered, data samples, and pre-fit simulated predictions, as well as some of the validations that were performed to ascertain that the analysis works as expected. The results of a joint ν+$\bar{\nu}$ oscillation fit with data amounting to 1.45×10^{21} protons on target at the far detector are presented and discussed in Chap. 6.

After the results have been presented, a short summary and discussion of the future prospects for neutrino oscillation physics in general and the T2K analysis in particular are given in Chap. 7.

In the process of developing the analysis presented in this thesis, two other oscillation analyses were performed using only antineutrino data: a measurement of the antineutrino oscillation parameters $\sin^2\bar{\theta}_{23}$ and $\Delta\bar{m}^2_{32}$ from $\bar{\nu}_\mu$ disappearance, and the first ever search for $\bar{\nu}_e$ appearance, using the antineutrino beam at T2K. Space does not permit a full discussion of these two analyses in the detail they deserve, but the methods and results are briefly presented in Appendix A. The $\bar{\nu}_\mu$ disappearance analysis has been published in Physical Review Letters [9], and both results have been summarised in conference proceedings such as [10–12].

Unfortunately, some references are made in this thesis to internal T2K technical notes, which may not be publicly available. Every effort has been made to reference articles in the public domain, but where this has not been possible information from the internal technical notes may be obtained by contacting the T2K Spokesperson and International Co-Spokesperson. Contact details can be found at http://t2k-experiment.org/contact-us/.

References

1. F. Close, *Antimatter* (OUP Oxford, 2009)
2. B. Kayser, Two questions about neutrinos (2010), arXiv:1012.4469 [hep-ph]
3. B. Kayser, Neutrino mass, mixing, and flavour change. J. Phys. G **33** (2006), http://pdg.lbl.gov/2007/reviews/numixrpp.pdf
4. K. Nakamura, S.T. Petcov, Neutrino mass, mixing, and oscillations. Rev. Part. Phys. Chin. Phys. C **38**, 235–258 (2014). (Updated June 2016), http://www-pdg.lbl.gov/2016/reviews/rpp2016-rev-neutrino-mixing.pdf
5. M. Fukugita, T. Yanagida, Baryogenesis without grand unification. Phys. Lett. B **174**(1), 45–47 (1986). doi:10.1016/0370-2693(86)91126-3
6. Z. Xing, S. Zhou, *Neutrinos in Particle Physics, Astronomy and Cosmology* (Zhejiang University Press, Hangzhou, 2011)
7. K. Abe et al. (T2K Collaboration), Measurements of neutrino oscillation in appearance and disappearance channels by the T2K experiment with 6.6×10^{20} protons on target. Phys. Rev. D **91**, 072010 (2015). doi:10.1103/PhysRevD.91.072010
8. K. Abe et al. (T2K Collaboration), Observation of electron neutrino appearance in a muon neutrino beam. Phys. Rev. Lett. **112**, 061802 (2014). doi:10.1103/PhysRevLett.112.061802
9. K. Abe et al. (T2K Collaboration), Measurement of muon antineutrino oscillations with an accelerator-produced off-axis beam. Phys. Rev. Lett. **116**, 181801 (2016). doi:10.1103/PhysRevLett.116.181801
10. M.R. Salzgeber (for the T2K Collaboration), Anti-neutrino oscillations with T2K, in *Proceedings of Science (EPS-HEP 2015)*, arXiv:1508.06153 [hep-ex]
11. K.E. Duffy (for the T2K Collaboration), Results and prospects from T2K, in *Proceedings from XVII International Workshop on Neutrino Factories and Future Neutrino Facilities (NuFact15)*, Rio de Janeiro vol. eConf C1508102 (2016), http://www.slac.stanford.edu/econf/C1508102/separado/2.pdf
12. K.E. Duffy (for the T2K Collaboration), First antineutrino oscillation results from T2K, in *Proceedings from NuPhys2015, Prospects in Neutrino Physics, London* (2015), arXiv:1605.01626 [hep-ex]

Chapter 2
Introduction to Neutrino Oscillation

The story of the neutrino begins in the early 20th century, when James Chadwick measured the energy spectrum of electrons released in beta decay to be continuous. In the contemporary understanding of beta decay as a two-body problem, this could only be explained by violating the law of conservation of energy (an explanation proposed by many). In late 1930, Wolfgang Pauli put forward a "desperate remedy" in a letter to Lise Meitner and the participants of a radioactivity conference in Tübingen (addressed "Liebe Radioaktive Damen und Herren"—"Dear Radioactive Ladies and Gentlemen") [1]: a light, electrically neutral, spin-$\frac{1}{2}$ particle that had somehow escaped all observation. The idea was developed almost immediately by Enrico Fermi as part of his theory of β-decay, published in 1934 [2], in which he gave Pauli's particle a name: the neutrino.[1]

In the most part, the original properties of the neutrino as predicted by Pauli and Fermi have been borne out through scientific testing in the many years since the theory was first proposed. We do indeed have evidence for a 'new' light electrically neutral fermion, which was first tentatively observed by Reines and Cowan in 1953 and confirmed in 1956 [3]. The main difference between the particle we know now and the particle that was described in Fermi's theory is its mass: Fermi's neutrino was massless, as indeed the neutrino is in the Standard Model of Particle Physics, but this is no longer believed to be true. The Nobel Prize was awarded in 2015 for measurements by the Super-Kamiokande and SNO experiments of neutrino flavour change, confirming that neutrinos have mass. Although the precise masses of the neutrinos are yet to be measured, it is considered a well-established fact that they are non-zero.

This chapter aims to give the reader an overview of the theory and experimental observations related to neutrino oscillation physics (the topic of this thesis). There are a wide range of other sources that are recommended for those seeking a deeper

[1]This is actually a shortcut in the story: Pauli named his particle the neutron, until the neutron was discovered and found not to be the particle that Pauli postulated. Edoardo Amaldi, in a conversation with Fermi, was the first to coin the word 'neutrino' for Pauli's particle, and the term was popularised by its use in Fermi's theory of beta decay.

K.E. Duffy, *First Measurement of Neutrino and Antineutrino Oscillation at T2K*,
Springer Theses, DOI 10.1007/978-3-319-65040-1_2

discussion of any of the topics in this chapter, which will be highlighted where possible. Historical accounts, written for the general public, of the discovery of the neutrino and the early days of neutrino physics can be found in [4, 5]. Griffiths [6] is a general particle physics textbook that covers the history of the field and puts the neutrino in context within the Standard Model of Particle Physics. For neutrino physics in particular, the reader is strongly recommended to read anything written by Boris Kayser, whose articles provide a well-written and understandable theoretical treatment of neutrino oscillation physics. The summer school lectures [7, 8], and review of neutrino mass, mixing, and flavour change in the 2007 Review of Particle Physics [9] are especially useful and relevant to the material covered here. The review of neutrino mass, mixing, and oscillation in the 2016 Review of Particle Physics [10] and Chap.4 of [11] are also excellent resources, giving a detailed overview of the current state of neutrino physics, from the theoretical framework to current experiments and measurements. Finally, there are a number of textbooks dedicated to the topic of neutrino physics, among which [12–14] are commonly recommended. There is, of course, always more to discuss, and for further references the reader is directed to the Resource Letter ANP-1: Advances in Neutrino Physics written by Maury Goodman [15]. This letter provides an excellent guide to the literature on the subject of neutrino oscillation, separated by topic, and indicates the approximate level of each resource.

Section 2.1 gives a brief overview of the history of neutrino physics, from the initial proposal and observation of the new particle to the measurement of two further neutrinos and the discovery that there can only be three types of light weakly-interacting neutrino. Section 2.2 introduces the phenomenon of neutrino oscillation, which is the subject of this thesis. The experimental evidence for oscillation is presented in Sect. 2.2.1, the theory behind it in Sect. 2.2.2, and the current state of world knowledge relating to neutrino oscillation is discussed in Sect. 2.2.3.

2.1 The Discovery of the Neutrino

Fermi's theory accurately accounted for almost all the observed properties of beta decay, and its success was taken as convincing evidence for the neutrino. It was soon noted that the theory predicted a reaction in which a free neutrino or antineutrino would interact in matter and be stopped, and could therefore be detected:

$$\bar{\nu} + p \rightarrow e^{+} + n \tag{2.1}$$

It would take another 22 years after the publication of Fermi's theory for technology to develop to a point that an observation of this reaction was possible, with the advent of very intense sources of neutrinos in the form of fission bombs and reactors. Frederick Reines and Clyde L. Cowan made the first direct measurement of the electron antineutrino via the reaction shown in Eq. 2.1, using a liquid scintillator detector positioned near a nuclear reactor (as described by Frederick Reines in [3] and his

Nobel Lecture [16]). The event signature was a prompt signal from e^+ annihilation in coincidence with a delayed signal from n capture on cadmium. The initial experiment [17], deployed in 1953 at the plutonium-producing nuclear reactor in Hanford, Washington, used a 300-litre detector filled with liquid scintillator: by far the largest particle detector built at the time. Unfortunately, the Hanford experiment suffered from a large cosmic ray background, which meant that although they saw hints of the presence of neutrinos they were not able to make a definitive measurement.

The experiment was redesigned with better segmentation (to allow for better separation of signals from cosmic rays and the reactor), and a larger volume (4200 litres of scintillator), and moved to the Savannah River Plant in South Carolina [18, 19]. Here, the detector could be situated underground, further reducing the cosmic-ray background. In June 1956, the Savannah River experiment confirmed the tentative findings of the Hanford experiment: they had detected antineutrinos.

A second experiment around the same time, conducted by Ray Davis [20], showed that antineutrinos from a reactor cannot produce electrons. Tanks containing carbon tetrachloride were placed near the Brookhaven nuclear reactor in an attempt to induce the reaction $\bar{\nu} + \mathrm{Cl} \rightarrow e^- + \mathrm{Ar}$, and no significant excess of Argon atoms was found. This result, taken together with Reines and Cowan's observation of antineutrinos producing positrons, led to the idea of conservation of lepton number (which is 1 for leptons and -1 for antileptons).

Following these first observations, advances in the field of experimental neutrino physics were rapid. In 1962 an experiment by Leon Lederman, Melvin Schwartz, Jack Steinberger, and collaborators [21] proved that a second (muon) neutrino existed and was paired with the muon the way the known (electron) neutrino was paired with the electron. 29 muon events and six electron events were measured from interactions in a ν_μ beam (with small ν_e contamination); if the muon and electron neutrinos were the same, the flavour symmetry of the weak interaction would predict equal numbers of events. This result was confirmed in two high-statistics measurements at CERN in 1963 [22, 23].

Even before the discovery of the muon neutrino, since the discovery of the muon in 1936, many had asked whether there were further heavier particles with similar interactions. A series of experiments performed between 1975 and 1977 by Martin L. Perl et al. [24] (see also Perl's Nobel Lecture [25]) using the SLC e^+e^- collider at SLAC answered this question with the discovery of a new lepton: the τ. By this time, the theory of the weak interaction had already been developed and the symmetry of the theory implied that if the τ lepton existed there must exist a full third generation of leptons, including the ν_τ. This was first observed many years later, in the year 2000, in an experiment performed by the DONuT collaboration at Fermi National Accelerator Laboratory [26]. τ leptons produced in ν_τ scattering were identified in plates of emulsion by the characteristic kinked track that comes from subsequent τ decay.

With the discovery of the τ, the next obvious question was: where does the pattern end? Are there only three generations of particles, or does it continue to higher energies and heavier leptons? As discussed in many particle physics textbooks such as [12], a partial answer to this question in the neutrino sector came from studies of

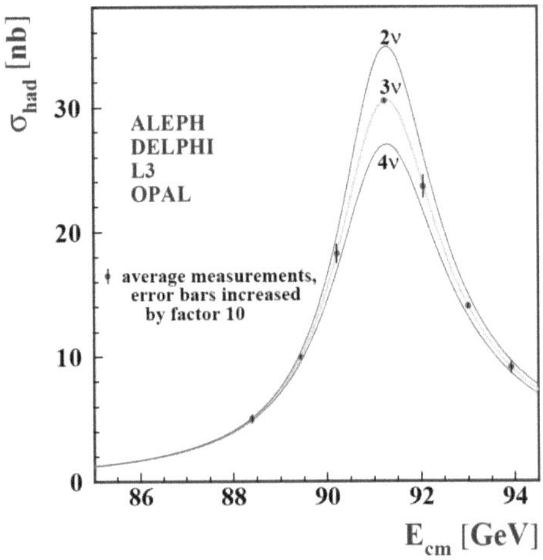

Fig. 2.1 The hadron production cross-section in e^+e^- collisions close to the Z resonance. The curves show the prediction for two, three, and four neutrino species in the Standard Model, and the points show the experimental data from a combination of the ALEPH, DELPHI, L3, and OPAL experiments at the LEP e^+e^- collider. Reprinted from Phys. Rep., vol. 427 (5–6), The ALEPH Collaboration et al., "Precision Electroweak Measurements on the Z Resonance", pp. 257–454, Copyright (2006), with permission from Elsevier [27]

Z production in e^+e^- collisions. The 'invisible' partial width of the Z-boson mass peak, Γ_{inv}, is determined by subtracting the measured 'visible' partial widths (from decays to hadrons and leptons) from the total width. This invisible width is assumed to be due to N_ν light neutrino species (i.e. neutrinos which are lighter than half the mass of the Z, so that the decay $Z \to \nu\bar{\nu}$ is kinematically allowed), each of which is assumed to contribute equally. A combined fit from the four experiments at the LEP electron-positron collider and the SLD experiment at the SLC e^+e^- collider gives $N_\nu = 2.984 \pm 0.008$ [27], consistent with three light neutrino species as shown in Fig. 2.1. This does not, however, mean there can only be three neutrino species: there may be more that do not couple to the Z (sterile neutrinos) or that it is impossible to produce in Z decay (heavy neutrinos).

2.2 Neutrino Oscillation

The focus of this thesis is neutrino oscillation, or neutrino flavour change as a function of energy and distance travelled by the particle, at T2K. There is very good evidence—from many sources—that neutrinos change from one flavour to another, some of which is summarised in Sect. 2.2.1. An overview of the theory and implications of neutrino oscillation is given in Sect. 2.2.2, and finally, Sect. 2.2.3 contains a review of the current state of world knowledge about the parameters governing neutrino oscillation.

2.2.1 *Initial Evidence for Neutrino Oscillation*

2.2.1.1 The Solar Neutrino Problem

In the late 1960s, a decade after his original neutrino experiment, Ray Davis and collaborators devised another experiment to measure solar neutrinos in the Homestake mine in South Dakota [28]. The aim was to measure neutrinos from the sun—particularly high-energy electron neutrinos produced by ^{8}B decay ($^{8}_{5}\mathrm{B} \rightarrow^{8}_{4}$ $\mathrm{Be} + e^{+} + \nu_e$)—using the same radiochemical method of detection as in Davis' previous experiment, $\nu_e + \mathrm{Cl} \rightarrow e^{-} + \mathrm{Ar}$. The results, published in 1968, came as a surprise to many: the upper bound on the solar ν_e flux was 2-3 times smaller than the prediction. This became known as the "solar neutrino problem".

Not everyone was surprised by this measurement. Bruno Pontecorvo had been working on a theory of massive neutrinos and neutrino oscillation [29], which predicted the solar neutrino problem. However, the success of the theory of the neutrino as a massless particle at the time meant that there was very little support for Pontecorvo's interpretation from the community. Most people believed that the solution to the problem would lie in the Standard Solar Model and the predicted neutrino flux, rather than the Standard Model of Particle Physics.

In 1989, the Kamiokande collaboration, using a large water Cherenkov detector originally intended to measure proton decay, produced a measurement of the solar neutrino flux that supported Davis' result [30]. Solar ν_e were detected via the recoil electron from elastic scattering ($\nu_e + e^{-} \rightarrow \nu_e + e^{-}$) and again they found the ratio of observed to predicted flux was around $\frac{1}{2}$. However, this measurement suffered from the same problems as Davis': it depended heavily on solar models to predict the neutrino flux, and could only measure high-energy neutrinos from ^{8}B interactions, which were not very well understood.

Additional evidence for ν_e disappearance came from the GALLEX [31] and SAGE [32] experiments in the early 1990s. As in Davis' experiment, both used radiochemical detection methods, but in this case with Gallium instead of Chlorine: $\nu_e +^{71}\mathrm{Ga} \rightarrow e^{-} +^{71}\mathrm{Ge}$. This interaction has a much lower threshold than $\nu_e + \mathrm{Cl}$, and so allowed the experiments to measure lower-energy solar neutrinos produced by different interactions in the sun. This was especially important because they were able to detect neutrinos produced by the solar pp-cycle, a chain of nuclear reactions starting with proton-proton fusion that was well known and understood at the time. Again, both experiments measured around half the predicted event rate. This was additional and stronger evidence for the solar neutrino problem, but still depended on fallible models to predict the neutrino flux from the sun.

The first model-independent measurement of the solar neutrino flux came from the SNO collaboration in 2002 [33]. As in Davis' experiment and Kamiokande, they measured mostly high-energy neutrinos from solar ^{8}B decay, but SNO measured both the ν_e flux (through charged-current interactions: $\nu_e + d \rightarrow p + p + e^{-}$, which only occur for ν_e) and the total neutrino flux (through flavour-independent neutral-current interactions, $\nu_x + d \rightarrow p + n + \nu_x$, and elastic scattering, $\nu_x + e^{-} \rightarrow \nu_x +$

e^-, where x could be e, μ, or τ). The measured total neutrino flux was in agreement with the Standard Solar Model, and the measured ratio of the ν_e flux to the total neutrino flux was 0.301 ± 0.033. This gave very strong evidence that ν_e were changing flavour into ν_μ and ν_τ in the sun.

An excellent overview of the solar neutrino problem and eventual solution is given in Arthur McDonald's Nobel Lecture [34].

2.2.1.2 Super-Kamiokande and Atmospheric Neutrinos

In 1998, the Super-Kamiokande collaboration (the successor to Kamiokande) published a measurement of atmospheric ν_μ disappearance [35], of which hints had been seen previously in the Kamiokande [36–38] and IMB [39, 40] water Cherenkov detectors. Super-Kamiokande has a good ability to measure the neutrino direction, and the flux of atmospheric ν_μ (produced by decays of cosmic-ray muons in the Earth's atmosphere) was measured as a function of incoming angle.

An accessible discussion of the Super-Kamiokande measurement and its impact on the field of neutrino physics is given in Takaaki Kajita's Nobel Lecutre [41]. Broadly speaking, if the atmospheric neutrinos are separated into two samples, up-going and down-going, it is possible to measure neutrino oscillations over two different baselines in the same detector. Down-going neutrinos will have a path length from the muon decay site of around 20–500 km, whereas up-going neutrinos (most of which will have passed through the Earth) have a path length from production of around 500–12,000 km. If we assume that cosmic rays (that produce neutrinos visible in Super-Kamiokande) are isotropic and that there is no ν_μ disappearance, Gauss's law predicts equal fluxes for the up-going and down-going neutrinos. However, the results from Super-Kamiokande showed an up-going neutrino flux of around half of the down-going flux, proving that the number of observed ν_μ depends on the distance between the neutrino production point in the atmosphere and the detector. Coming 4 years before the SNO result, this was the first model-independent evidence for neutrino flavour change.

Figure 2.2 shows the ratio of measured to predicted number of events as a function of distance travelled by the neutrino, L (calculated from the angle of the incoming neutrino), and neutrino energy, E_ν, in Super-Kamiokande. This is shown for both electron-like and muon-like signals, corresponding to ν_e and ν_μ interactions respectively. At high L/E_ν, the observed ν_μ flux is around 50% of the prediction—clear evidence for ν_μ disappearance.

2.2.1.3 KamLAND and Reactor Neutrinos

The KamLAND experiment ran from 2002–2004 and provided very strong evidence for the oscillation of $\bar{\nu}_e$ produced in nuclear reactors [42–44]. The experiment measured antineutrinos produced in 55 Japanese nuclear power reactors, with an average baseline between the antineutrino production point and the detector of 180 km. Not

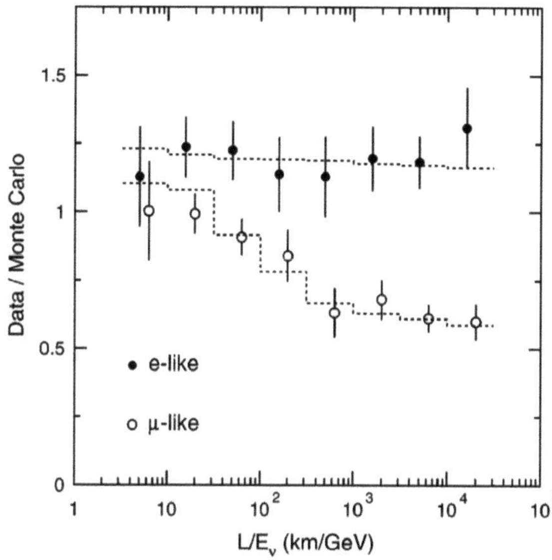

Fig. 2.2 Ratio of number of data events to predicted number of events without neutrino oscillations (from Monte Carlo) as a function of L/E$_\nu$ in Super-Kamiokande. The points show the ratio of data to prediction, and the dashed lines show the expected shape when accounting for $\nu_\mu \rightarrow \nu_\tau$ oscillation in a two-flavour oscillation model with $\Delta m^2 = 2.2 \times 10^{-3}$eV^2 and $\sin^2 2\theta = 1$. Reprinted figure with permission from Y. Fukuda et al. (Super-Kamiokande Collaboration), Phys. Rev. Lett., vol. 81, pp. 1562–1567 (1998). Copyright (1998) by the American Physical Society [35]

only did they observe fewer $\bar{\nu}_e$ than expected, but they were also able to show that the probability to measure a $\bar{\nu}_e$ oscillated as a function of 1/E$_\nu$ as in Fig. 2.3, which is consistent with what we expect from the PMNS model prediction (discussed in Sect. 2.2.2).

Since these first decades, many other experiments have provided improved measurements and stronger evidence for neutrino oscillation; [10] and [11] provide a good overview of the current experimental evidence and efforts. The evidence is now overwhelmingly strong that neutrino oscillations do occur, and the focus in the field has moved into precise determination of the theoretical parameters. The current state of knowledge about the parameters governing neutrino oscillation is discussed in Sect. 2.2.3.

2.2.2 *The Theory of Neutrino Oscillation*

The charged-current weak interaction (as far as we know) always couples a neutrino of a given flavour to a charged lepton of the same flavour. For example, a decay of a W boson may produce an electron along with an electron neutrino, or a muon neutrino can exchange a W boson with a neutron to produce a muon and a proton. In this way, we can define the neutrino flavour eigenstates by their associated charged

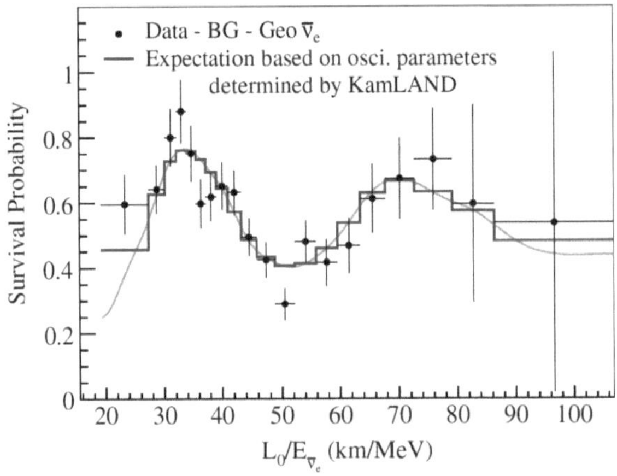

Fig. 2.3 Ratio of number of $\bar{\nu}_e$ data events observed at KamLAND (with background and geoneutrino contributions subtracted) to the expectation in the case of no oscillation, as a function of L_0/E_ν, where L_0 is the effective baseline taken from a flux-weighted average across all reactors, $L_0 = 180$ km. Reprinted figure with permission from S. Abe et al. (KamLAND Collaboration), Phys. Rev. Lett., vol. 100, p. 221803 (2008). Copyright (2008) by the American Physical Society [44]

leptons in W interactions: a decay which produces a lepton l_α ($\alpha = e, \mu, \tau$) will by definition also produce a neutrino of the same flavour, ν_α.

The fact that neutrinos have mass means that there must exist some spectrum of neutrino mass eigenstates ν_i ($i = 1, 2, 3, ...$), each with well-defined mass m_i (at this point we will assume that neutrinos have mass, but later we will see that the theory of neutrino oscillation relies on it, so the observations of neutrino oscillation described above are equivalent to observations that neutrinos have non-zero mass). Because the flavour and mass eigenstates each form a complete set, the flavour states must be linear superpositions of these mass eigenstates:

$$|\nu_\alpha\rangle = \sum_i \mathrm{U}^*_{\alpha i}|\nu_i\rangle \tag{2.2}$$

Equation 2.2 defines the leptonic mixing matrix, U, commonly known as the PMNS (Pontecorvo-Maki-Nakagawa-Sakata) matrix, which relates the mass and flavour eigenstates. In the case of three mass eigenstates and three flavour eigenstates, U takes the form:

$$\mathrm{U} = \begin{pmatrix} \mathrm{U}_{e1} & \mathrm{U}_{e2} & \mathrm{U}_{e3} \\ \mathrm{U}_{\mu 1} & \mathrm{U}_{\mu 2} & \mathrm{U}_{\mu 3} \\ \mathrm{U}_{\tau 1} & \mathrm{U}_{\tau 2} & \mathrm{U}_{\tau 3} \end{pmatrix}$$

such that the 'e' row of U gives the linear combination of neutrino mass eigenstates that couple to the electron, and the '1' column of U gives the linear combination of flavour eigenstates that constitute the mass eigenstate ν_1.

Further discussion of the PMNS formalism can be found in many references, such as [7, 11]. The 3×3 PMNS matrix can be parameterised in terms of three mixing angles θ_{12}, θ_{13}, and θ_{23} and three CP-violating phases δ_{CP}, α_1, and α_2:

$$U = \underbrace{\begin{pmatrix} 1 & 0 & 0 \\ 0 & c_{23} & s_{23} \\ 0 & -s_{23} & c_{23} \end{pmatrix}}_{Atmospheric} \underbrace{\begin{pmatrix} c_{13} & 0 & s_{13}e^{-i\delta_{CP}} \\ 0 & 1 & 0 \\ -s_{13}e^{i\delta_{CP}} & 0 & c_{13} \end{pmatrix}}_{Cross-mixing} \underbrace{\begin{pmatrix} c_{12} & s_{12} & 0 \\ -s_{12} & c_{12} & 0 \\ 0 & 0 & 1 \end{pmatrix}}_{Solar} \underbrace{\begin{pmatrix} e^{i\alpha_1/2} & 0 & 0 \\ 0 & e^{i\alpha_2/2} & 0 \\ 0 & 0 & 1 \end{pmatrix}}_{Majorana} \quad (2.3)$$

where $c_{ij} \equiv \cos\theta_{ij}$ and $s_{ij} \equiv \sin\theta_{ij}$.

U has been decomposed into four component matrices in Eq. 2.3 to make it easier to read and interpret, and because different mixing angles are measured by different types of experiment. The first matrix includes only the mixing angle θ_{23}, which is responsible for the majority of the oscillation seen in atmospheric neutrinos. It is approximately equal to the atmospheric mixing angle θ_{atm} measured when the oscillation of atmospheric ν_μ into ν_τ is approximated by a two-flavour neutrino oscillation (neglecting ν_e). The third matrix includes only the mixing angle θ_{12}, which dominates the mixing of solar neutrinos—if we approximate the oscillation of solar ν_e into ν_μ in a two-flavour model with mixing anlge θ_{sol}, θ_{sol} is very close to θ_{12}.

The second matrix is known as the cross-mixing matrix, and depends on two parameters: the mixing angle θ_{13} and the CP-violating phase δ_{CP}. As we shall see later, a non-zero δ_{CP} will lead to a complex matrix U and different probabilities for the CP-conjugate oscillations $\nu_\alpha \rightarrow \nu_\beta$ and $\bar{\nu}_\alpha \rightarrow \bar{\nu}_\beta$ (which would be a significant finding). Note that δ_{CP} only appears in terms where it is multiplied by $\sin\theta_{13}$, meaning that our ability to measure a difference in the probabilities $P(\nu_\alpha \rightarrow \nu_\beta)$ and $P(\bar{\nu}_\alpha \rightarrow \bar{\nu}_\beta)$ is dependent on the size of $\sin\theta_{13}$ (in particular, it requires $\sin\theta_{13} \neq 0$). However, the mixing angle θ_{13} is not special—if we multiply out the matrices, we find that we require all three mixing angles to be non-zero to measure CP violation. The reason U is usually separated in this way is to emphasize the reliance on θ_{13}, as this is the smallest mixing angle and the last to be measured. For this reason, confirming that $\theta_{13} \neq 0$ was a major goal in neutrino physics in recent years. The first evidence for $\theta_{13} \neq 0$ at 5σ came from the the Daya Bay Neutrino Experiment in 2012 [45], followed by a 7.3σ confirmation from T2K in 2014 [46].

The final matrix in Eq. 2.3 contains the so-called "Majorana" CP-violating phases. These lead to physical effects only for the case of Majorana neutrinos (i.e. where neutrinos are their own antiparticle), and do not conserve lepton number. Even in the case of Majorana neutrinos, these CP-violating phases do not affect the oscillation probability, which (as in Eq. 2.11) depends on $\sum_{i>j} U^*_{\alpha i} U_{\alpha j}$, meaning that the Majorana phases on the diagonal of the matrix cancel. It is not possible to determine whether neutrinos are their own antiparticles from oscillation measurements; other experiments (such as the search for neutrinoless double beta decay being conducted or planned by the MAJORANA [47], GERDA [48], CUORE [49], and SNO+ [50] collaborations, among others) are needed to answer this question. Discussions of the potential implications of, and experimental searches for, Majorana neutrinos can be found in [7, 9, 11, 51].

In the Standard Model, the only charged lepton ν_α couples to is l_α, so the three ν_α flavour states must be orthogonal. If $\delta_{\alpha\beta}$ represents the Kronecker delta:

$$\begin{aligned}\langle\nu_\alpha|\nu_\beta\rangle &= \langle\sum_i \mathrm{U}^*_{\alpha i}\nu_i|\sum_j \mathrm{U}^*_{\beta j}\nu_j\rangle \\ \delta_{\alpha\beta} &= \sum_{i,j} \mathrm{U}_{\alpha i}\mathrm{U}^*_{\beta j}\langle\nu_i|\nu_j\rangle \\ \delta_{\alpha\beta} &= \sum_i \mathrm{U}_{\alpha i}\mathrm{U}^*_{\beta j}\end{aligned} \tag{2.4}$$

where the final equivalence uses the assumption that the mass eigenstates are orthogonal (this will be true unless some masses m_i, m_j are degenerate). This means that U must be unitary.

In order to make up the three known orthogonal ν_α, we require at least three ν_i states. However, the total number of flavour or mass eigenstates is not known. This means that although U must be unitary, the same is not necessarily true for the 3×3 sub-matrix given above.

We can invert Eq. 2.2 to write the mass eigenstates as superpositions of the flavour eigenstates:

$$|\nu_i\rangle = \sum_\alpha \mathrm{U}_{\alpha i}|\nu_\alpha\rangle$$

and therefore calculate the flavour-α fraction of each mass eigenstate ν_i to be $|\mathrm{U}_{\alpha i}|^2$. This can be interpreted as the probability of measuring a charged lepton of flavour α from a ν_i interaction. Figure 2.4 shows the flavour composition of each of the mass eigenstates (assuming there are only three flavour and three mass states), according to current measurements. The left-hand side of the diagram shows the case assuming normal mass hierarchy (NH, $m_3^2 > m_2^2$), and the right-hand side shows the case for inverted mass hierarchy (IH, $m_2^2 > m_3^2$). It is not yet known which mass hierarchy is true in nature, although we do know—from measurements of neutrino oscillation in the matter of the sun as discussed in [9]—that $m_2^2 > m_1^2$.

Figure 2.5 shows an example schematic of neutrino oscillation. At the source, a neutrino is produced together with a charged lepton of flavour α: this is, by defintition, a ν_α. It then travels a distance L to a detector where it interacts with some target to produce a second charged lepton of flavour β. At the point it interacted this neutrino was, by definition, a ν_β, and if $\beta \neq \alpha$ the neutrino has changed flavour.

The following derivation of the probability for neutrino flavour change follows the treatment found in many lecture series and textbooks, such as [6–8, 12–14]. Since, as in Eq. 2.2, we can write the flavour states as a sum over mass eigenstates, we can write the amplitude for the process $\nu_\alpha \to \nu_\beta$, $\mathcal{A}(\nu_\alpha \to \nu_\beta)$, as a sum over all mass

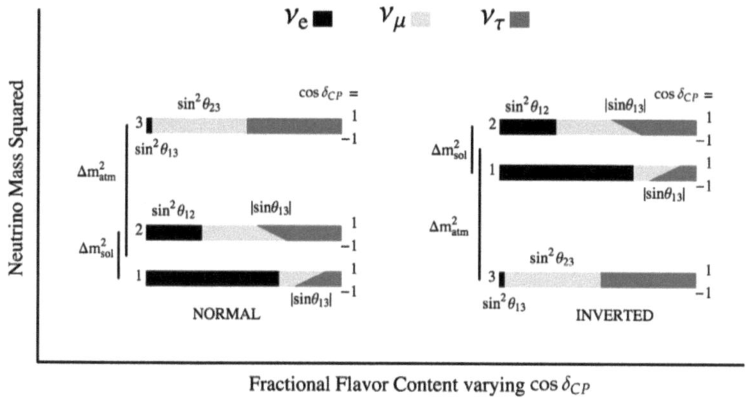

Fig. 2.4 The neutrino mass eigenstates ν_1, ν_2, and ν_3 (labelled simply '1', '2', and '3'), showing the flavour fraction of each mass eigenstate according to measurements of solar, reactor, atmospheric, and long-baseline accelerator neutrinos. The width of the lines is used to show how the fractions change as $\cos\delta_{CP}$ changes from -1 to $+1$. Both normal and inverted mass hierarchies are shown. Adapted figure with permission from O. Mena and S. Parke, Phys. Rev. D, vol. 69, p. 117301 (2004). Copyright (2004) by the American Physical Society [53]

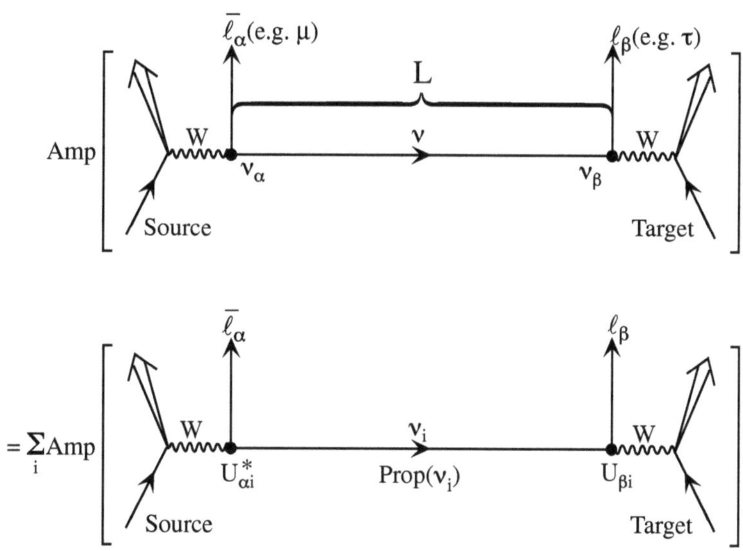

Fig. 2.5 A schematic showing neutrino flavour change in a vacuum. 'Amp' denotes the amplitude, $\mathcal{A}$, and 'Prop(ν_i)' denotes the propagator, $\mathcal{P}_i$. Figure from [7]

eigenstates. Each mass eigenstate will propagate according to some propagator $\mathcal{P}_i = \langle \nu_i(0)|\nu_i(x_i)\rangle = \exp(-ip_i \cdot x_i)$, where p_i and x_i are the momentum and position four-vectors of the mass eigenstate. Therefore we can write:

$$\begin{aligned}\mathcal{A}(\nu_\alpha \to \nu_\beta) &= \langle \nu_\beta | \nu_\alpha(L) \rangle \\ &= \sum_i \mathrm{U}_{\alpha i} \langle \nu_\beta | \nu_i(L) \rangle \\ &= \sum_i \mathrm{U}_{\alpha i} \mathcal{P}_i \langle \nu_\beta | \nu_i(0) \rangle \\ &= \sum_i \sum_\gamma \mathrm{U}_{\alpha i} \mathrm{U}^*_{\gamma i} \mathcal{P}_i \langle \nu_\beta | \nu_\gamma \rangle \\ &= \sum_i \mathrm{U}_{\alpha i} \mathrm{U}^*_{\beta i} \mathcal{P}_i \end{aligned} \tag{2.5}$$

In the lab frame, $p_i \cdot x_i = E_i t - |\mathbf{p}_i| L$, where t is the time taken for the neutrino to travel a distance L and $\mathbf{p}_i$ is the momentum three-vector (note that the subscript i here refers to the eigenstate ν_i and not a component of the three-vector). Since we know that neutrino masses are very small, $m_i^2 \ll E_i^2$ for any realistic energy E_i, we can approximate the three-momentum by:

$$|\mathbf{p}_i| = \sqrt{E_i^2 - m_i^2} \simeq E_i - \frac{m_i^2}{2E_i} \tag{2.6}$$

from which we find that

$$p_i \cdot x_i \simeq E_i(t - L) + \frac{m_i^2}{2E_i} L \tag{2.7}$$

If we make the assumption that all mass-eigenstates ν_i composing the initial flavour state ν_α have the same energy E, then Eq. 2.7 becomes

$$p_i \cdot x_i \simeq E(t - L) + \frac{m_i^2}{2E} L \tag{2.8}$$

We can factor out the phase $E(t - L)$ for the rest of this calculation because it is common to all mass eigenstates and therefore irrelevant for neutrino oscillation. In this case:

$$\mathcal{P}_i \propto \exp(-i \frac{m_i^2}{2E} L) \tag{2.9}$$

and the amplitude for a neutrino to change from a ν_α into a ν_β, from Eq. 2.5, is:

$$\mathcal{A}(\nu_\alpha \to \nu_\beta) \propto \sum_i \mathrm{U}^*_{\alpha i} \mathrm{U}_{\beta i} e^{-i m_i^2 \frac{L}{2E}} \tag{2.10}$$

Squaring this, we find that the probability $P(\nu_\alpha \to \nu_\beta)$ for $\nu_\alpha \to \nu_\beta$ is:

$$\begin{aligned} P(\nu_\alpha \to \nu_\beta) &= |\mathcal{A}(\nu_\alpha \to \nu_\beta)|^2 \\ &= \delta_{\alpha\beta} - 4\sum_{i>j} \Re(\mathrm{U}^*_{\alpha i}\mathrm{U}_{\beta i}\mathrm{U}_{\alpha j}\mathrm{U}^*_{\beta j}) \sin^2(\Delta m^2_{ij}\frac{L}{4E}) \\ &+ 2\sum_{i>j} \Im(\mathrm{U}^*_{\alpha i}\mathrm{U}_{\beta i}\mathrm{U}_{\alpha j}\mathrm{U}^*_{\beta j}) \sin(\Delta m^2_{ij}\frac{L}{2E}) \end{aligned} \tag{2.11}$$

where

$$\Delta m^2_{ij} \equiv m^2_i - m^2_j$$

At this point the reader might well point out an unsatisfactory element of derivation: the same-energy assumption made between Eqs. 2.7 and 2.8 was made without justification and may be questionable. In fact, we have no reason to believe that all neutrino mass eigenstates are emitted from the source with the same energy (indeed, by examining the simple kinematics of particles emitted in 2-body decays it is easy to show that we expect particles with different masses to have different energies and momenta). The full, more correct treatment of neutrino oscillations is more involved than we need for our purposes and contains many quantum-mechanical subtleties that are still under debate. However, all agree on the final result. This derivation may be overly simplistic but contains all the essential physics and arrives at the correct experession for the oscillation probability (given in Eq. 2.11). A more correct treatment and discussion of why this incorrect treatment nevertheless arrives at the same result is given in [54].

We can make a number of statements based on Eq. 2.11:

- When $P(\nu_\alpha \to \nu_\beta)$ is written in this way it is immediately obvious that neutrino flavour change can only occur if neutrinos have some non-zero mass. If the neutrino masses are all zero, $\Delta m^2_{ij} = 0$ for all i and j. In this case, Eq. 2.11 reduces to $P(\nu_\alpha \to \nu_\beta) = \delta_{\alpha\beta}$ so no flavour change can take place. For this reason, measurements of neutrino flavour change are equivalent to showing that neutrinos have mass.
- The probability for a neutrino to change flavour oscillates as a function of L/E. This is why the phenomenon is commonly referred to as "neutrino oscillation".
- Using the fact that U is unitary, it can be easily shown that $\sum_\beta P(\nu_\alpha \to \nu_\beta) = 1$. The total neutrino flux remains constant; it is just redistributed among the flavours.
- The oscillation probability for antineutrinos, $P(\bar{\nu}_\alpha \to \bar{\nu}_\beta)$, can be written as [7]:

$$\begin{aligned} P(\bar{\nu}_\alpha \to \bar{\nu}_\beta) &= \delta_{\alpha\beta} - 4\sum_{i>j} \Re(\mathrm{U}^*_{\alpha i}\mathrm{U}_{\beta i}\mathrm{U}_{\alpha j}\mathrm{U}^*_{\beta j}) \sin^2(\Delta m^2_{ij}\frac{L}{4E}) \\ &- 2\sum_{i>j} \Im(\mathrm{U}^*_{\alpha i}\mathrm{U}_{\beta i}\mathrm{U}_{\alpha j}\mathrm{U}^*_{\beta j}) \sin(\Delta m^2_{ij}\frac{L}{2E}) \end{aligned} \tag{2.12}$$

Note that the only difference between Eq. 2.12 and the CP-conjugate Eq. 2.11 is the sign of the third, imaginary, term. Therefore, if U is complex $P(\nu_\alpha \rightarrow \nu_\beta) \neq P(\bar{\nu}_\alpha \rightarrow \bar{\nu}_\beta)$, which violates CP symmetry. Conversely, if U is real there will be no CP violation in neutrino oscillation. The derivation of this antineutrino oscillation probability assumes that CPT symmetry always holds, because the same oscillation parameters are used to describe the oscillation of neutrinos and antineutrinos.

- The oscillation probability depends on the mass-squared splittings, $\Delta m^2_{ij} \equiv m^2_i - m^2_j$, not the neutrino masses directly. This means that measurements of neutrino oscillation can give us information about the differences between the masses of the different mass eigenstates, but not their absolute scale (i.e. how far the whole pattern lies above zero).
- When we include the factors of $\hbar$ and c that have been set equal to 1 so far, the argument of the $\sin^2$ term becomes

$$\Delta m^2_{ij}(\mathrm{eV}^2)\frac{1.27L(\mathrm{km})}{E(\mathrm{GeV})}$$

$\sin^2(x)$ is substantial for $x \sim \mathcal{O}(1)$ or larger, which means that an experiment with a given L and E will be sensitive to mass-squared splittings $\sim (1.27L/E)^{-1}$ or larger. T2K, which has a baseline of 295 km and a peak neutrino energy of 0.6 GeV, is therefore sensitive to mass-squared splittings down to $\mathcal{O}(10^{-3}\mathrm{eV}^2)$. This means that—according to current measurements, detailed in Sect. 2.2.3—T2K is sensitive to oscillations according to the mass-squared splitting Δm^2_{32}, but not Δm^2_{21}.

T2K measures neutrino oscillation in two channels: disappearance of ν_μ (or $\bar{\nu}_\mu$) and appearance of ν_e (or $\bar{\nu}_e$). To first order, the oscillation probabilities are given by:

$$\begin{aligned} P(\nu_\mu(\bar{\nu}_\mu) \rightarrow \nu_\mu(\bar{\nu}_\mu)) &\simeq 1 - 4\cos^2\theta_{13}\sin^2\theta_{23}[1 - \cos^2\theta_{13}\sin^2\theta_{23}]\sin^2\frac{\Delta m^2_{32}L}{4E} \\ &+ [\text{solar term, matter effect term}] \end{aligned} \tag{2.13}$$

$$\begin{aligned} P(\nu_\mu(\bar{\nu}_\mu) \rightarrow \nu_e(\bar{\nu}_e)) &\simeq \sin^2\theta_{23}\sin^2 2\theta_{13}\sin^2\frac{\Delta m^2_{31}L}{4E} \\ &+ \sin 2\theta_{12}\sin 2\theta_{23}\sin 2\theta_{13}\cos\theta_{13}\sin\frac{\Delta m^2_{21}L}{4E}\sin\frac{\Delta m^2_{31}L}{4E} \\ &\times [\cos\frac{\Delta m^2_{32}L}{4E}\cos\delta_{CP} - (+)\sin\frac{\Delta m^2_{32}L}{4E}\sin\delta_{CP}] \\ &+ [\text{solar term, matter effect term}] \end{aligned} \tag{2.14}$$

where brackets are used to indicate the antineutrino oscillation probabilities. The neutrino and antineutrino disappearance probabilities, shown in Eq. 2.13, are identical (barring matter effects—see Sect. 2.2.2.1), and the neutrino and antineutrino appearance probabilities, shown in Eq. 2.14, have opposite signs in the third term.

This is the term which allows for CP violation. The so-called 'solar' terms (terms in which the oscillation is governed by the solar mass-squared splitting, Δm^2_{21}) have been omitted in Eqs. 2.13 and 2.14 because T2K is not sensitive to these oscillations. The full oscillation probability, including solar terms and matter effect corrections, is used for all calculations presented in this thesis.

2.2.2.1 Neutrino Oscillation in Matter

In the calculations above we have not considered the medium through which the neutrinos travel (we have calculated the probability for neutrino oscillation in a vacuum). In real experiments, neutrinos travel through matter—often the crust of the Earth—and for certain experiments this may have a significant effect on the oscillation physics.

There are two possible interactions between neutrinos and matter particles in the earth, shown in Fig. 2.6: charged-current scattering of a ν_e from an electron via exchange of a W boson, or neutral-current scattering of any flavour of neutrino from an electron, proton, or neutron via exchange of a Z boson. Both interactions effectively create additional potentials that neutrinos experience as they travel through matter, which scale with the number of particles in the matter.

Because neutral-current scattering is independent of neutrino flavour, the additional potential from these interactions with matter does not affect the oscillation physics. However, the potential due to charged-current scattering affects only ν_e, and so will have an effect on each of the mass eigenstates ν_1, ν_2, and ν_3 according to their ν_e component.

The exact mechanism by which this affects the oscillation of neutrinos is too detailed for repetition here, but is described well in many sources such as [7, 9, 14, 55]. For this thesis it will suffice to say that the matter effect can change neutrino oscillation probabilities, and must be taken into account in oscillation measurements.

A significant implication when considering the oscillation of antineutrinos, as in this thesis, is that the sign of the additional potential due to charged-current scattering reverses for antineutrinos compared to neutrinos. This is because the matter through

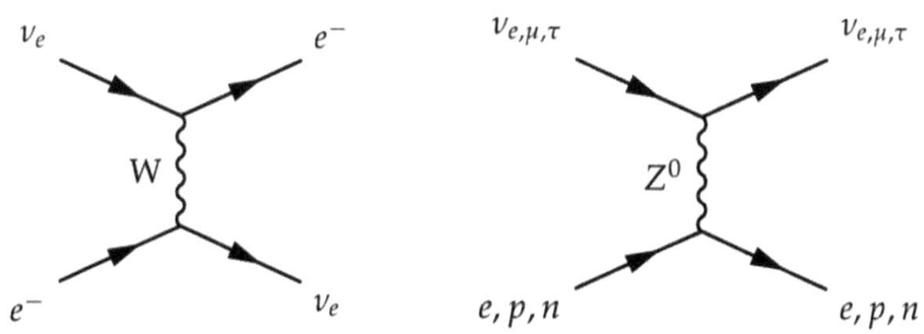

Fig. 2.6 Feynman diagrams for charged-current (*left*) and neutral-current (*right*) neutrino scattering in matter

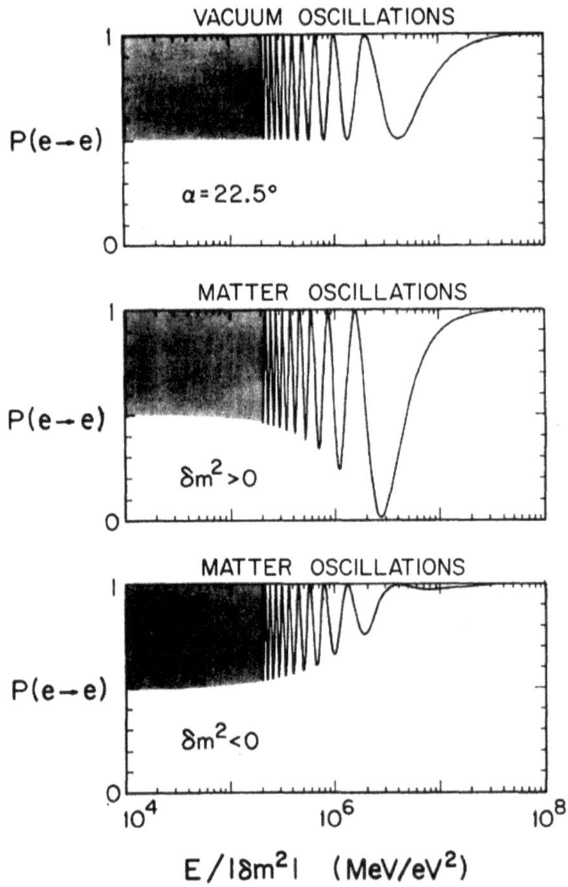

Fig. 2.7 The ν_e survival probability in a two-neutrino model over a baseline L = 5×10^3 km, as a function of $E/|\Delta m^2|$. *Top* $P(\nu_e \to \nu_e)$ prediction in vacuum; *middle* $P(\nu_e \to \nu_e)$ prediction in matter given $\Delta m^2 > 0$; bottom $P(\nu_e \to \nu_e)$ prediction in matter given $\Delta m^2 < 0$. This demonstrates the effect of the sign of Δm^2 on the neutrino oscillation probability in matter. Reprinted figure with permission from V. Barger, K. Whisnant, S. Pakvasa, and R. J. N. Phillips, Phys. Rev. D, vol. 22, pp. 2718–2726 (1980). Copyright (1980) by the American Physical Society [52]

which the neutrino is travelling is made up of particles rather than antiparticles, and therefore is itself CP-asymmetric. This leads to a "fake CP violation", which can cause us to measure $P(\bar{\nu}_\alpha \to \bar{\nu}_\beta) \neq P(\nu_\alpha \to \nu_\beta)$ but does not tell us anything about CP violation in lepton interactions at the fundamental level. Because of this it is very important to account for matter effects correctly in searches for genuine neutrino-sector CP violation.

It can also be shown [52] that the neutrino oscillation probability in matter, unlike in vacuum, *is* dependent on the sign of the mass splittings Δm^2_{ij}. This principle is demonstrated in Fig. 2.7, where the probability of electron neutrino survival is plotted for some assumed oscillation parameters and a baseline L = 5×10^3 km, assuming a two-neutrino (ν_e and ν_μ) model. The top plot shows the predicted oscillation probability $P(\nu_e \to \nu_e)$ in vacuum, the middle plot shows the prediction in matter given that $\Delta m^2 > 0$, and the bottom plot shows the prediction for $\Delta m^2 < 0$. The matter effect in this figure is much more significant than in any realistic long-baseline neutrino oscillation experiment, and the generalisation to the three-neutrino case is

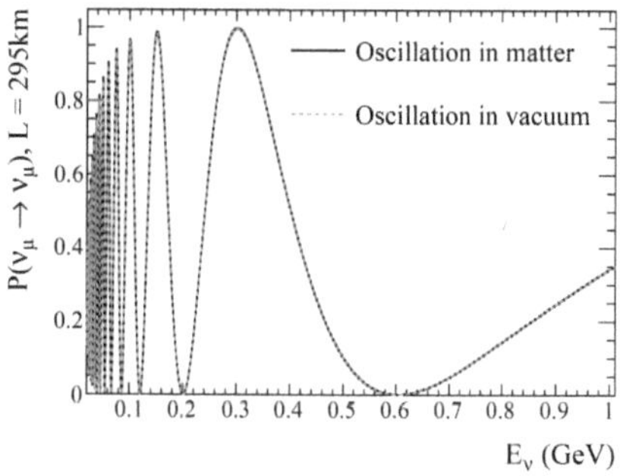

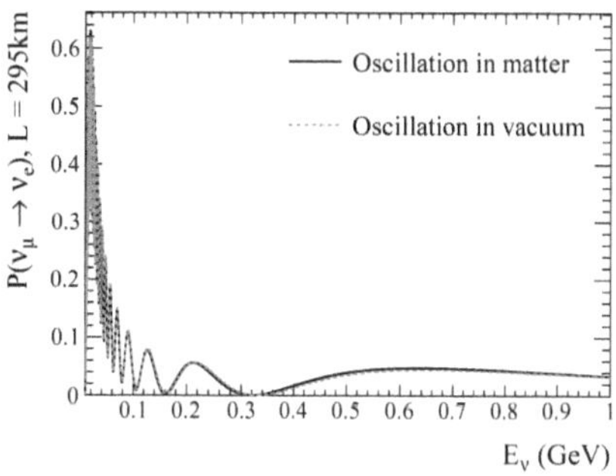

Fig. 2.8 Oscillation probabilities for ν_μ disappearance (*left*) and ν_e appearance (*right*) at T2K, calculated in vacuum and matter. The following oscillation parameters were assumed: $\sin^2 2\theta_{12} = 0.846$, $\sin^2 2\theta_{13} = 0.085$, $\sin^2 2\theta_{23} = 1.0$, $\Delta m^2_{12} = 7.53 \times 10^{-5}\text{eV}^2$, $\Delta m^2_{32} = 2.5 \times 10^{-3}\text{eV}^2$, $\delta_{CP} = 0$, as well as a constant earth density of 2.6g/cm^3. Figure produced using the Prob3++ software package [57]

more complicated. However, the result still stands: the oscillation probability in matter depends on the sign of Δm^2. This means that it may be possible to determine the mass hierarchy with a long-baseline neutrino experiment.

The T2K baseline and energy are such that matter effects are very small. The average matter density along the baseline between J-PARC and Super-Kamiokande is 2.6 g/cm^3 [56]. Figure 2.8 shows the oscillation probability as a function of energy for the T2K baseline in matter (black solid line) and in vacuum (grey dashed line). It is difficult to distinguish between the two lines, indicating that the matter effect is not significant for T2K. Despite this, the matter effect is taken into account in all calculations of the neutrino oscillation probabilities in this thesis.

2.2.3 *Current World Knowledge and Open Questions*

A large number of measurements of neutrino oscillation now exist from solar, atmospheric, reactor, and long- and short-baseline accelerator neutrino experiments. Taken together, the results can give us good estimates of many of the neutrino oscillation parameters. The current world-best results before the analysis described in this thesis (as determined by the Particle Data Group in the 2015 update to the 2014 Review of Particle Physics [58]) are summarised below.

- $\boldsymbol{\theta_{12}}$:
  The best measurement of the solar mixing angle, θ_{12}, comes from a three-neutrino oscillation fit to KamLAND and global solar neutrino data, using constraints on θ_{13} from accelerator and short-baseline reactor neutrino experiments [59]. They report

$$\tan^2\theta_{12} = 0.436^{+0.029}_{-0.025}$$

which corresponds to

$$\sin^2\theta_{12} = 0.304^{+0.014}_{-0.013}$$

- $\mathbf{\Delta m^2_{21}}$:
The current world-leading measurement of Δm^2_{21} comes from the same KamLAND and global solar neutrino fit as the measurement of θ_{12} [59], which finds

$$\Delta m^2_{21} = (7.53 \pm 0.18) \times 10^{-5}\text{eV}^2$$

- $\boldsymbol{\theta_{23}}$ **and** $\mathbf{\Delta m^2_{32}}$:
Most experiments present results as a two-dimensional contour in $\sin^2\theta_{23}$—Δm^2_{32} space because these two parameters are highly correlated. This is then projected onto one axis to give an estimate with 1D uncertainties for each parameter. Figure 2.9 shows the world-leading contours in $\sin^2\theta_{23}$–Δm^2_{32} space, from T2K, Super-Kamiokande (SK), and MINOS [60]. T2K contours are shown for both 68 and 90% confidence regions, in the case of both normal and inverted mass hierarchies. SK and MINOS contours are shown for 90% confidence regions, assuming normal hierarchy only. The projections on the top and right hand side of the plot show the 1D T2K $\Delta\chi^2$ maps for $\sin^2\theta_{23}$ and Δm^2_{32} respectively, from which the T2K 1D best-fit points and uncertainties can be inferred.

T2K currently has the world-leading measurement of $\sin^2\theta_{23}$, which depends on the mass hierarchy:

$$\begin{aligned}\sin^2\theta_{23} &= 0.514^{+0.055}_{-0.056}\ \text{(NH)}\\ &= 0.511^{+0.055}_{-0.055}\ \text{(IH)}\end{aligned}$$

The Particle Data Group gives a global best-fit value for $|\Delta m^2_{32}|$ using data from the T2K [60], MINOS [62], and Daya Bay [63] experiments:

$$\begin{aligned}|\Delta m^2_{32}| &= (2.42 \pm 0.06) \times 10^{-3}\text{eV}^2\ \text{(NH)}\\ &= (2.49 \pm 0.06) \times 10^{-3}\text{eV}^2\ \text{(IH)}\end{aligned}$$

- $\boldsymbol{\theta_{13}}$:
The most precise measurement of θ_{13} comes from the reactor experiments Daya Bay [63, 64], Double Chooz [65, 66], and RENO [67]. A global fit to this data gives:

$$\sin^2 2\theta_{13} = (8.5 \pm 0.5) \times 10^{-2}$$

which corresponds to

$$\sin^2\theta_{13} = (2.19 \pm 0.12) \times 10^{-2}$$

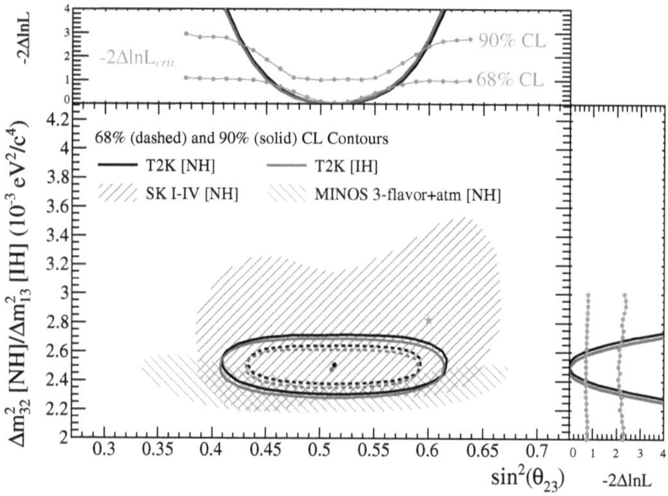

Fig. 2.9 68 and 90% confidence regions for $\sin^2\theta_{23}$ and Δm^2_{32}(NH) or Δm^2_{13}(IH) from T2K, overlaid with 90% confidence regions from the Super-Kamiokande [61] and MINOS [62] collaborations for NH. The 1D profiled likelihoods from T2K are shown at the top and the right of the plot, with lines overlaid showing the points representing the 1D critical values of $-2\Delta ln\mathcal{L}$ at 68 and 90%. Reprinted figure with permission from K. Abe et al. (T2K Collaboration), Phys. Rev. Lett., vol. 112, p. 181801 (2014). Copyright (2014) by the American Physical Society [60]

- $\delta_{\mathbf{CP}}$:
  There are currently no measurements of δ_{CP}, although some hints to its value exist.

A T2K fit to both muon neutrino disappearance and electron neutrino appearance [68], using a Gaussian constraint on $\sin^2 2\theta_{13}$ from reactor neutrino experiments as given in the 2013 PDG [69] (the "reactor constraint"), give 68 and 90% credible intervals for δ_{CP} as shown in Fig. 2.10. The posterior probability density is maximum at $\delta_{CP} \simeq -\pi/2$.

Other hints come from the MINOS and NOνA collaborations. Figure 2.11 shows the significance for exclusion of δ_{CP} from a ν_e appearance fit at NOνA, using the reactor constraint [70]. The data mildly disfavour $\delta_{CP} \simeq \pi/2$, in agreement with the hints from T2K (which are presented on a scale of $(-\pi)$–π rather than 0–2π as shown by NOνA and MINOS).

Figure 2.12 shows the 1D likelihood profile from a fit to MINOS data including ν-dominated beam data, $\bar{\nu}$-dominated beam data, atmospheric neutrino data and the reactor constraint on $\sin^2 2\theta_{13}$ [62]. The 1D likelihood is given for a number of assumptions about the mass hierarchy and octant of θ_{23}, and overall they disfavour 36% of the δ_{CP} parameter space at 68% confidence level.

Interestingly, the values disfavoured by MINOS correspond to the values best favoured by T2K and NOνA. This highlights the fact that none of these results are statistically significant: we do not yet have a conclusive measurement of δ_{CP}.

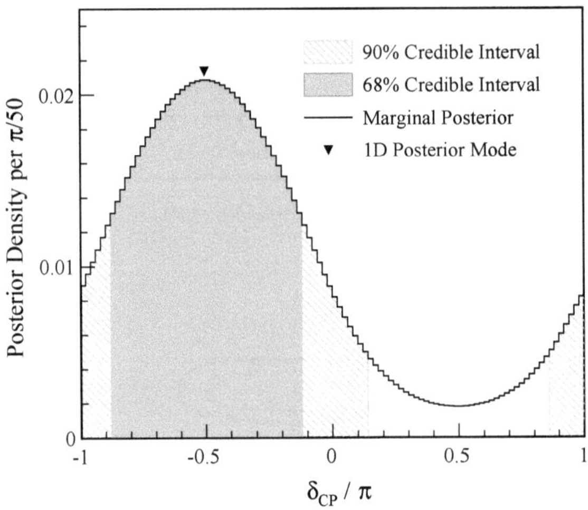

Fig. 2.10 1D posterior probability density in δ_{CP} from fit to T2K ν_μ disappearance and ν_e appearance data, with constraint on $\sin^2 2\theta_{13}$ from reactor experiments. Figure from [68]

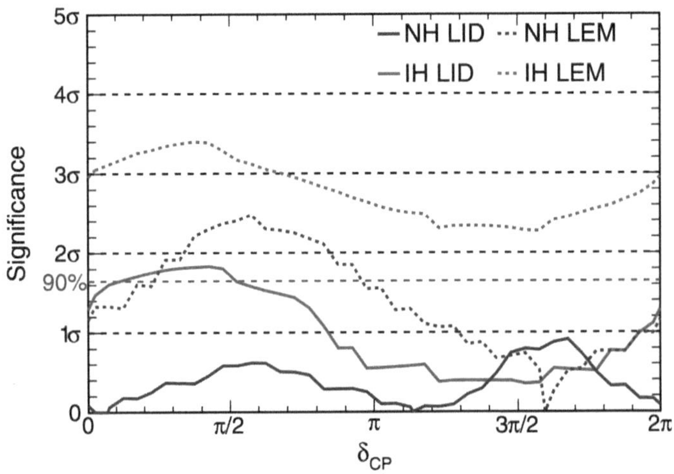

Fig. 2.11 1D significance as a function of δ_{CP} and mass hierarchy from the first NOνA ν_e appearance data, with constraint on $\sin^2 2\theta_{13}$ from reactor experiments. Results using two different event selections are shown: the primary (secondary) selection technique is shown with solid (dotted) lines. Using the primary selection, the data mildly disfavour $\delta_{CP} \in [0, 0.8\pi]$ in the inverted mass hierarchy ($> 1\sigma$). The normal hierarchy shows better consistency with the data in both selections. Reprinted figure with permission from P. Adamson et al. (NOνA Collaboration), Phys. Rev. Lett., vol. 116, p. 151806 (2016). Copyright (2016) by the American Physical Society [70]

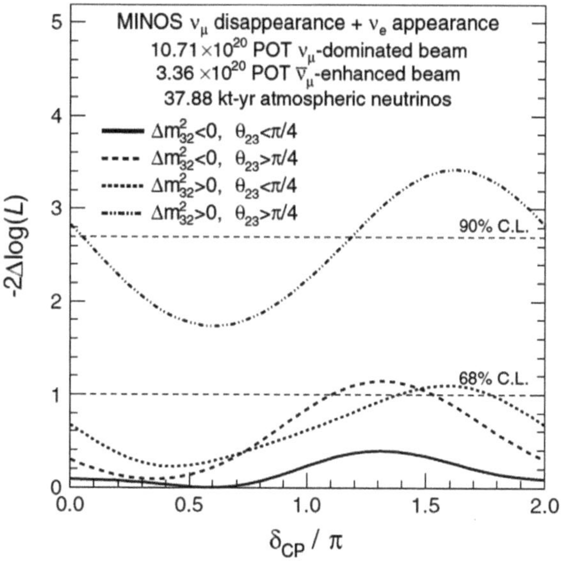

Fig. 2.12 1D significance as a function of δ_{CP} for each combination of mass hierarchy and θ_{23} octant from fits to MINOS beam ν, beam $\bar{\nu}$, and atmospheric ν data, with constraint on $\sin^2 2\theta_{13}$ from reactor experiments. The data mildly disfavour $\delta_{CP} \simeq 3\pi/2$, in disagreement with the hints from T2K and NOνA in Figs. 2.10 and 2.11. Reprinted figure with permission from P. Adamson et al. (MINOS Collaboration), Phys. Rev. Lett., vol. 112, p. 191801 (2014). Copyright (2014) by the American Physical Society [62]

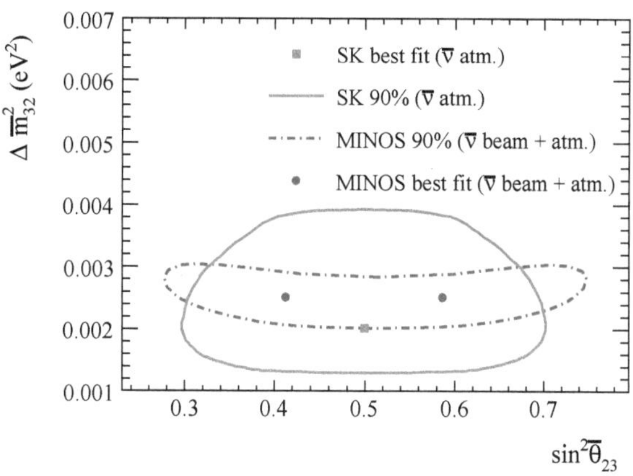

Fig. 2.13 90% confidence level contours and best-fit points in $\sin^2\bar{\theta}_{23}$–$\Delta\bar{m}^2_{32}$ from antineutrino measurements by the Super-Kamiokande [71] and MINOS [72] collaborations. The MINOS fit result was originally presented in terms of $\sin^2 2\bar{\theta}_{23}$ and has been transformed to $\sin^2\theta_{23}$; hence the two best-fit points

All of the results above have been produced by analyses assuming CPT symmetry (i.e. that the oscillation parameters that govern neutrino oscillation are the same as the ones that govern antineutrino oscillation), and that CP-asymmetry enters only through non-zero δ_{CP}. However, fits have also been done in which different parameters are assigned to neutrino and antineutrino oscillation. Disagreement between these two sets of oscillation parameters could indicate CP violation (if $P(\bar{\nu}_\alpha \to \bar{\nu}_\beta) \neq P(\nu_\alpha \to \nu_\beta)$) or CPT violation (if $P(\bar{\nu}_\alpha \to \bar{\nu}_\alpha) \neq P(\nu_\alpha \to \nu_\alpha)$) in the neutrino sector.

Figure 2.13 shows the current measurements of the parameters that govern $\bar{\nu}_\mu$ disappearance, $\sin^2 \bar{\theta}_{23}$ and $\Delta\bar{m}^2_{32}$, from the MINOS and Super-Kamiokande experiments. Both results are in agreement with the current measurements of the neutrino oscillation parameters $\sin^2 \theta_{23}$ and Δm^2_{32} and show no indication of CPT violation.

Measuring $\sin^2 \bar{\theta}_{23}$ and $\Delta\bar{m}^2_{32}$ at T2K is one of the subjects of this thesis, presented in Appendix A, Sect. A.3. Figure A.4 shows the T2K contour compared to those from MINOS and Super-Kamiokande: the results of all three experiments are in agreement with each other and with current estimates of the neutrino oscillation parameters.

References

1. W. Pauli, Letter of the 4th December 1930, Pauli Archive at CERN, http://microboone-docdb.fnal.gov/cgi-bin/RetrieveFile?docid=953;filename=pauli%20letter1930.pdf
2. E. Fermi, Versuch Einer Theorie der β-Strahlen: I. Zeitschrift für Physik **88**(3–4), 161–177 (1934). doi:10.1007/BF01351864
3. F. Reines, The early days of experimental neutrino physics. Science **203**(4375), 11–16 (1979). doi:10.1126/science.203.4375.11
4. F. Reines, C. Cowan, The reines-cowan experiments: detecting the poltergeist. Los Alamos Sci. **25**, 4–27 (1997)
5. F. Close, *Neutrino*. (OUP Oxford, 2010)
6. D. Griffiths, *Introduction to Elementary Particles*, 2nd, revised edn. (Wiley, 2008)
7. B. Kayser, Neutrino physics, eConf **C040802**, L004 (2004), arXiv:hep-ph/0506165
8. B. Kayser, Neutrino oscillation physics, in *Proceedings, 2011 European School of High-Energy Physics (ESHEP 2011): Cheile Gradistei*, Romania, 7–20 September 2011, pp. 107–117 (2014), doi:10.5170/CERN-2014-003.107, also available at arXiv:1206.4325 [hep-ph]
9. B. Kayser, Neutrino mass, mixing, and flavour change. J. Phys. G **33** (2006), http://pdg.lbl.gov/2007/reviews/numixrpp.pdf
10. K. Nakamura and S. T. Petcov, Neutrino mass, mixing, and oscillations. Rev. Part. Phys. Chin. Phys. C **38**, 235–258 ((2014) Updated June 2016), http://www-pdg.lbl.gov/2016/reviews/rpp2016-rev-neutrino-mixing.pdf
11. Fundamental physics at the intensity frontier: report of the workshop held December 2011 in Rockville, MD (2012), doi:10.2172/1042577, also available at arXiv:1205.2671 [hep-ex]
12. K. Zuber, *Neutrino Physics* 2nd edn. (Taylor & Francis, 2011)
13. V. Barger, D. Marfatia, K. Whisnant, *The Physics of Neutrinos* (Princeton Univ, Pr, 2012)
14. C. Giunti, C.W. Kim, *Fundamentals of Neutrino Physics and Astrophysics*. (OUP Oxford, 2007)
15. M.C. Goodman, Resource letter ANP-1: advances in neutrino physics. Am. J. Phys. **84**(12), 907–916 (2016). doi:10.1119/1.4962228
16. F. Reines, The neutrino: from poltergeist to particle. Rev. Mod. Phys. **68**, 317–327 (1996). doi:10.1103/RevModPhys.68.317, http://www.nobelprize.org/nobel_prizes/physics/laureates/1995/reines-lecture.html
17. F. Reines, C.L. Cowan, Detection of the free neutrino. Phys. Rev. **92**, 830–831 (1953). doi:10.1103/PhysRev.92.830
18. C.L. Cowan et al., Detection of the free neutrino: a confirmation. Science **124**(3212), 103–104 (1956). doi:10.1126/science.124.3212.103

19. F. Reines et al., Detection of the free antineutrino. Phys. Rev. **117**, 159–173 (1960). doi:10.1103/PhysRev.117.159
20. R. Davis, Attempt to detect the antineutrinos from a nuclear reactor by the $Cl^{37}(\bar{\nu}, e^-)A^{37}$ reaction. Phys. Rev. **97**, 766–769 (1955). doi:10.1103/PhysRev.97.766
21. G. Danby et al., Observation of high-energy neutrino reactions and the existence of two kinds of neutrinos. Phys. Rev. Lett. **9**, 36–44 (1962). doi:10.1103/PhysRevLett.9.36
22. M.M. Block et al., Neutrino interactions in the CERN heavy liquid bubble chamber. Phys. Lett. **12**(3), 281–285 (1964). doi:10.1016/0031-9163(64)91104-7
23. J.K. Bienlein et al., Spark chamber study of high-energy neutrino interactions. Phys. Lett. **13**(1), 80–86 (1964). doi:10.1016/0031-9163(64)90316-6
24. M.L. Perl et al., Evidence for anomalous lepton production in $e^+ - e^-$ annihilation. Phys. Rev. Lett. **35**, 1489–1492 (1975). doi:10.1103/PhysRevLett.35.1489
25. M.L. Perl, Reflections on the discovery of the tau lepton, nobel lectures in physics 1991–1995 (1997), https://www.nobelprize.org/nobel_prizes/physics/laureates/1995/perl-lecture.html
26. K. Kodama et al., Observation of tau neutrino interactions. Phys. Lett. B **504**(3), 218–224 (2001). doi:10.1016/S0370-2693(01)00307-0
27. The ALEPH Collaboration et al., Precision electroweak measurements on the Z resonance. Phys. Rep. **427**(5-6), 257–454 (2006). doi:10.1016/j.physrep.2005.12.006
28. R. Davis, D.S. Harmer, K.C. Hoffman, Search for neutrinos from the sun. Phys. Rev. Lett. **20**, 1205–1209 (1968). doi:10.1103/PhysRevLett.20.1205
29. B. Pontecorvo, Neutrino experiments and the problem of conservation of leptonic charge. J. Exp. Theor. Phys. **26**, 984–988 (1968), http://jetp.ac.ru/cgi-bin/e/index/e/26/5/p984?a=list
30. K.S. Hirata et al., Observation of ^8B solar neutrinos in the Kamiokande-II detector. Phys. Rev. Lett. **63**, 16–19 (1989). doi:10.1103/PhysRevLett.63.16
31. M. Cribier et al., Results of the whole gallex experiment. Nucl. Phys. B Proc. Suppl. **70**(1–3), 284–291 (1999). doi:10.1016/S0920-5632(98)00438-1, *Proceedings of the Fifth International Workshop on topics in Astroparticle and Underground Physics*
32. J.N. Abdurashitov et al., Solar neutrino flux measurements by the Soviet-American Gallium Experiment (SAGE) for half the 22-year solar cycle. J. Exp. Theor. Phys. **95**(2), 181–193 (2002). doi:10.1134/1.1506424
33. Q.R. Ahmad et al., (SNO Collaboration), Direct evidence for neutrino flavor transformation from neutral-current interactions in the sudbury neutrino observatory. Phys. Rev. Lett. **89**, 011301 (2002). doi:10.1103/PhysRevLett.89.011301
34. A.B. McDonald, The sudbury neutrino observatory: observation of flavor change for solar neutrinos. Rev. Mod. Phys. **88**(3), 030502 (2016). doi:10.1103/RevModPhys.88.030502, https://www.nobelprize.org/nobel_prizes/physics/laureates/2015/mcdonald-lecture.html
35. Y. Fukuda et al., (Super-Kamiokande Collaboration), Evidence for oscillation of atmospheric neutrinos. Phys. Rev. Lett. **81**, 1562–1567 (1998). doi:10.1103/PhysRevLett.81.1562
36. K.S. Hirata et al., Experimental study of the atmospheric neutrino flux. Phys. Lett. B **205**(2), 416–420 (1988). doi:10.1016/0370-2693(88)91690-5
37. K.S. Hirata et al., Observation of a small atmospheric ν_μ/ν_e ratio in Kamiokande. Phys. Lett. B **280**(1), 146–152 (1992). doi:10.1016/0370-2693(92)90788-6
38. Y. Fukuda et al., Atmospheric ν_μ/ν_e ratio in the Multi-GeV energy range. Phys. Lett. B **335**(2), 237–245 (1994). doi:10.1016/0370-2693(94)91420-6
39. D. Casper et al., Measurement of atmospheric neutrino composition with the IMB-3 detector. Phys. Rev. Lett. **66**, 2561–2564 (1991). doi:10.1103/PhysRevLett.66.2561
40. R. Becker-Szendy et al., Electron-and muon-neutrino content of the atmospheric flux. Phys. Rev. D **46**, 3720–3724 (1992). doi:10.1103/PhysRevD.46.3720
41. T. Kajita, Discovery of atmospheric neutrino oscillations. Rev. Mod. Phys. **88**(3), 030501 (2016). doi:10.1103/RevModPhys.88.030501, https://www.nobelprize.org/nobel_prizes/physics/laureates/2015/kajita-lecture.html
42. K. Eguchi et al., (KamLAND Collaboration), First results from kamLAND: evidence for reactor antineutrino disappearance. Phys. Rev. Lett. **90**, 021802 (2003). doi:10.1103/PhysRevLett.90.021802

43. T. Araki et al., (KamLAND Collaboration), Measurement of neutrino oscillation with KamLAND: evidence of spectral distortion. Phys. Rev. Lett. **94**, 081801 (2005). doi:10.1103/PhysRevLett.94.081801
44. S. Abe et al., (KamLAND Collaboration), Precision measurement of neutrino oscillation parameters with KamLAND. Phys. Rev. Lett. **100**, 221803 (2008). doi:10.1103/PhysRevLett.100.221803
45. F.P. An et al., (Daya Bay Collaboration), Observation of electron-antineutrino disappearance at daya bay. Phys. Rev. Lett. **108**, 171803 (2012). doi:10.1103/PhysRevLett.108.171803
46. K. Abe et al., (T2K Collaboration), Observation of electron neutrino appearance in a muon neutrino beam. Phys. Rev. Lett. **112**, 061802 (2014). doi:10.1103/PhysRevLett.112.061802
47. W. Xu et al. (Majorana Collaboration), The majorana demonstrator: a search for neutrinoless double-beta decay of 76 Ge. J. Phys. Conf. Ser. **606**(1), 012004 (2015), http://stacks.iop.org/1742-6596/606/i=1/a=012004
48. M. Agostini et al. (GERDA Collaboration), $2\nu\beta\beta$ Decay of 76 Ge Into Excited States With GERDA Phase I. J. Phys. G: Nucl. Part. Phys. **42**(11), 115201 (2015), http://stacks.iop.org/0954-3899/42/i=11/a=115201
49. C. Arnaboldi et al., (CUORE Collaboration), CUORE: a cryogenic underground observatory for rare events. Nucl. Instrum. Meth. Phys. Res. Sect. A **518**(3), 775–798 (2004). doi:10.1016/j.nima.2003.07.067
50. S. Andringa, E. Arushanova, S. Asahi et al., Current status and future prospects of the SNO+ Experiment. Adv. High Energy Phys. **2016**, (2016). doi:10.1155/2016/6194250
51. B. Kayser, Two questions about neutrinos (2010), arXiv:1012.4469 [hep-ph]
52. V. Barger, K. Whisnant, S. Pakvasa, R.J.N. Phillips, Matter effects on three-neutrino oscillations. Phys. Rev. D **22**, 2718–2726 (1980). doi:10.1103/PhysRevD.22.2718
53. O. Mena, S. Parke, Unified graphical summary of neutrino mixing parameters. Phys. Rev. D **69**, 117301 (2004). doi:10.1103/PhysRevD.69.117301
54. E. Kh. Akhmedov, A.Yu. Smirnov, Paradoxes of neutrino oscillations. Phys. At. Nucl. **72**(8), 1363–1381 (2009). doi:10.1134/S1063778809080122
55. E. Kh Akhmedov, Neutrino physics, lectures given at trieste summer school in particle physics, June 7–July 9, 1999 (2000), arXiv:hep-ph/0001264
56. K. Hagiwara, N. Okamura, K. Senda, The earth matter effects in neutrino oscillation experiments from Tokai to Kamioka and Korea. J. High Energy Phys. **2011**(9), 82 (2011). doi:10.1007/JHEP09(2011)082
57. http://www.phy.duke.edu/raw22/public/Prob3++/
58. K.A. Olive et al. (Particle Data Group), Review of particle physics. Chin. Phys. C **38**, 090001 ((2014) and 2015 update). doi:10.1088/1674-1137/38/9/090001
59. A. Gando et al., (KamLAND Collaboration), Reactor on-off antineutrino measurement with kamLAND. Phys. Rev. D **88**, 033001 (2013). doi:10.1103/PhysRevD.88.033001
60. K. Abe et al., (T2K Collaboration), Precise measurement of the neutrino mixing parameter θ_{23} from muon neutrino disappearance in an off-axis beam. Phys. Rev. Lett. **112**, 181801 (2014). doi:10.1103/PhysRevLett.112.181801
61. A. Himmel, Super-Kamiokande Collaboration, recent results from Super-Kamiokande. AIP Conf. Proc. **1604**(1), 345–352 (2014). doi:10.1063/1.4883450
62. P. Adamson et al., (MINOS Collaboration), Combined analysis of ν_μ disappearance and $\nu_\mu \to \nu_e$ appearance in MINOS using accelerator and atmospheric neutrinos. Phys. Rev. Lett. **112**, 191801 (2014). doi:10.1103/PhysRevLett.112.191801
63. F.P. An et al., (Daya Bay Collaboration), New measurement of antineutrino oscillation with the full detector configuration at daya bay. Phys. Rev. Lett. **115**, 111802 (2015). doi:10.1103/PhysRevLett.115.111802
64. F.P. An et al., (Daya Bay Collaboration), Independent measurement of the neutrino mixing angle θ_{13} via neutron capture on hydrogen at daya bay. Phys. Rev. D **90**, 071101 (2014). doi:10.1103/PhysRevD.90.071101
65. Y. Abe et al., (Double Chooz Collaboration), Ortho-Positronium observation in the double chooz experiment. J. High Energy Phys. **2014**(10), 32 (2014). doi:10.1007/JHEP10(2014)032

66. Y. Abe et al., (Double Chooz Collaboration), First Measurement of θ_{13} from delayed neutron capture on hydrogen in the double chooz experiment. Phys. Lett. B **723**(1–3), 66–70 (2013). doi:10.1016/j.physletb.2013.04.050
67. J.K. Ahn et al., (RENO Collaboration), Observation of reactor electron antineutrinos disappearance in the RENO experiment. Phys. Rev. Lett. **108**, 191802 (2012). doi:10.1103/PhysRevLett.108.191802
68. K. Abe et al., (T2K Collaboration), Measurements of neutrino oscillation in appearance and disappearance channels by the T2K experiment With 6.6×10^{20} protons on target. Phys. Rev. D **91**, 072010 (2015). doi:10.1103/PhysRevD.91.072010
69. J. Beringer et al. (Particle Data Group), Review of particle physics. Phys. Rev. D **86**, 010001 (2012) and (2013 partial update for the 2014 edition). doi:10.1103/PhysRevD.86.010001
70. P. Adamson et al., (NOvA Collaboration), First measurement of electron neutrino appearance in NOvA. Phys. Rev. Lett. **116**, 151806 (2016). doi:10.1103/PhysRevLett.116.151806
71. K. Abe et al., (Super-Kamiokande Collaboration), Search for differences in oscillation parameters for atmospheric neutrinos and antineutrinos at Super-Kamiokande. Phys. Rev. Lett. **107**, 241801 (2011). doi:10.1103/PhysRevLett.107.241801
72. P. Adamson et al., (MINOS Collaboration), Measurement of neutrino and antineutrino oscillations using beam and atmospheric data in MINOS. Phys. Rev. Lett. **110**, 251801 (2013). doi:10.1103/PhysRevLett.110.251801

Chapter 3
The T2K Experiment

T2K (Fig. 3.1, described in more detail in [1]) is a long-baseline neutrino oscillation experiment designed to observe and measure the mixing of muon neutrinos with the other neutrino flavours. It uses an intense beam of muon neutrinos created by a 30 GeV proton beam at the J-PARC facility in Tokai, Japan. The proton beam is directed onto a graphite target, and the resulting hadrons focussed by magnetic horns (which can select either π^+, for a beam composed mainly of ν_μ, or π^-, for a beam composed mainly of $\bar{\nu}_\mu$) into a 96 m tunnel, where they decay to produce neutrinos. The neutrino beam is measured by two near detectors 280 m from the target, and a far detector, Super-Kamiokande.

The baseline between the neutrino production point and the far detector, 295 km, is carefully chosen to correspond to the first minimum in the ν_μ survival probability at the peak neutrino beam energy of 0.6 GeV (as shown in Fig. 3.5). This allows T2K to measure neutrino oscillation in two channels: disappearance of ν_μ (for which the probability is dominated by the oscillation parameters $\sin^2\theta_{23}$ and Δm^2_{32}), and appearance of ν_e (which is also sensitive to the oscillation parameters $\sin^2\theta_{13}$ and δ_{CP}). At the chosen neutrino energy and baseline, T2K is not sensitive to the other two oscillation parameters, $\sin^2\theta_{12}$ and Δm^2_{21}. ν_τ appearance in the ν_μ beam also occurs, but charged-current ν_τ interactions are not measured because it is extremely unlikely for τ leptons to be produced at the neutrino energies of the T2K beam (the threshold energy for producing a τ in charged-current neutrino scattering on a neutron is around 3.5 GeV, and the oscillation probability at such high energies is small). Neutral-current ν_τ interactions are accounted for by calculating the neutral-current interaction rate according to the unoscillated neutrino flux.

In order to measure the oscillation parameters at T2K, the expected observables at the far detector are predicted based on neutrino beam and neutrino-nucleus interaction models. Measurements from both near detectors are used to estimate the model parameters and associated uncertainties, to reduce the uncertainty on the far detector prediction.

The T2K beam, including the beamline and neutrino flux simulation, is described in Sect. 3.1. The two near detectors, INGRID and ND280, are discussed in Sects. 3.2.1

K.E. Duffy, *First Measurement of Neutrino and Antineutrino Oscillation at T2K*,
Springer Theses, DOI 10.1007/978-3-319-65040-1_3

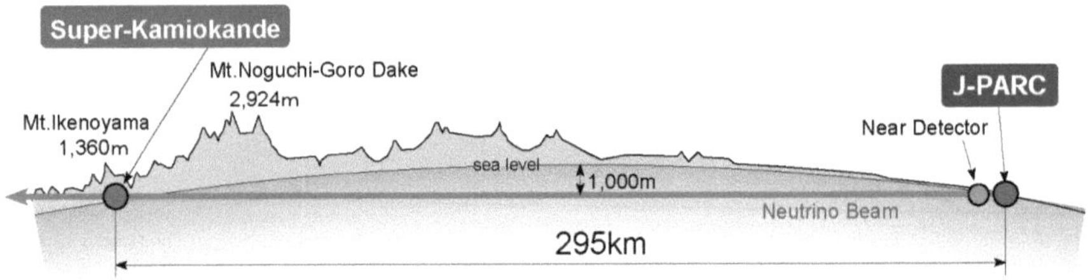

Fig. 3.1 The T2K (Tokai-to-Kamioka) experiment. Copyright by the T2K Collaboration

and 3.2.2 respectively, and the far detector, Super-Kamiokande, is described in Sect. 3.3. A short discussion of neutrino interactions in T2K is given in Sect. 3.4, and the models used for simulation of neutrino interactions in Sect. 3.4.1. The subject of neutrino interactions is a broad and rich area of current research, and beyond the scope of this thesis. However, understanding interaction cross sections is vital to measuring neutrino oscillation, and the limits of our understanding form an important systematic uncertainty on the oscillation results. Finally, a summary of the data collected at T2K and analysed in this thesis is given in Sect. 3.5.

3.1 The T2K Beam

The J-PARC beamline [1] (which was newly constructed for T2K in Tokai) consists of three accelerators: a linear accelerator (LINAC), a rapid-cycling synchrotron (RCS), and a main ring (MR).

Particles are first accelerated in the LINAC, where an H^- beam is accelerated up to 400 MeV. This is converted into a beam of H^+ by charge-stripping foils as it is injected into the RCS, where it is accelerated up to 3 GeV. The RCS runs with a 25 Hz cycle and two bunches per cycle. Around 5% of these bunches are supplied to the MR, where they are accelerated up to 30 GeV. The rest are supplied to muon and neutron beamlines for other experiments at J-PARC.

To produce the T2K beam, the proton beam is extracted from the MR within a single turn by a set of 5 kicker magnets. A single extraction of the proton beam is known as a 'spill': each spill contains 8 proton bunches of less than 100 ns each, separated by around 560 ns, and has a total duration of around 5 μs. Knowing the time structure of the extracted proton beam is very important for the T2K beam trigger—this makes it possible to discriminate various backgrounds (including cosmic rays) from the beam signal in the T2K detectors.

3.1.1 *The T2K Neutrino Beamline*

The 8 proton bunches extracted from the MR in a single spill are directed into the T2K neutrino beamline, which consists of two sections: the primary and secondary beamline, as shown in Fig. 3.2. In the primary beamline, the extracted proton beam is bent to point in the direction of the secondary beamline (towards Kamioka), and focussed to have the desired profile at the target. The intensity, position, profile, and loss of the proton beam are precisely measured in the primary beamline, because a well-tuned proton beam is essential for stable neutrino beam production.

As it enters the secondary beamline, the proton beam impinges on a graphite target to produce secondary mesons (mostly pions and kaons), which are focussed by magnetic horns before decaying to produce neutrinos. The target is a graphite rod of length 91.4 cm (equivalent to 1.9 interaction lengths for the proton beam) and diameter 2.6 cm. It is cooled by a helium gas flow to counteract the heat load from the pulsed beam.

The secondary mesons are focussed by three magnetic horns [3]. The target is situated inside the first horn, which collects together the mesons produced by proton interactions, and the second and third horns focus these mesons into a beam. Each magnetic horn consists of two coaxial conductors that form a closed volume. A toroidal magnetic field is generated in the volume between the conductors, which varies as 1/(distance from horn axis). The horns are operated with a 250 kA pulsed current, and produce a maximum magnetic field of 1.7 T. The neutrino flux at the far

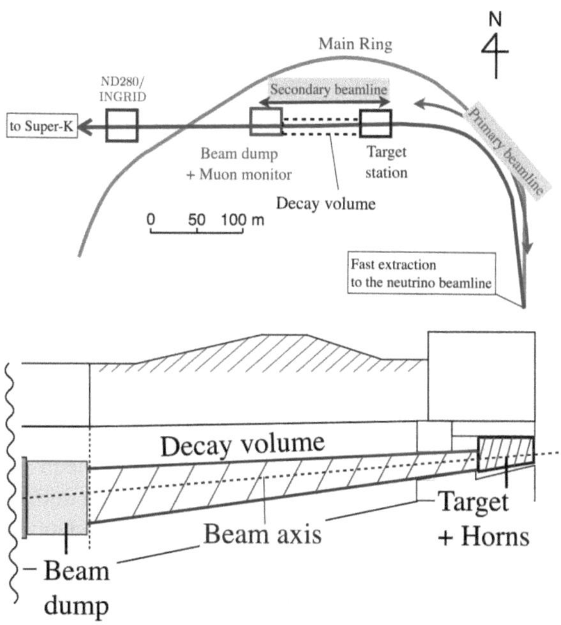

Fig. 3.2 *Top* overview of the T2K beamline. *Bottom* side view of the secondary beam line, showing the target, magnetic horns, decay volume and beam dump. Figure from [2]

detector is increased by a factor of ~17 at the spectrum peak energy by the use of these magnetic horns.

The current driving the magnetic horns can be inverted to focus either positively- or negatively-charged mesons, resulting in a predominantly neutrino or antineutrino beam. The current used to focus positively charged mesons and produce a beam which is mostly composed of neutrinos, is referred to as 'Forward Horn Current' (FHC). The current used to focus negatively charged mesons and produce a predominantly antineutrino beam is referred to as 'Reverse Horn Current' (RHC). The terms FHC and RHC are used interchangeably with 'neutrino-mode beam' and 'antineutrino-mode beam' in this thesis.

The mesons which are focussed by the magnetic horns enter a 96 m long steel tunnel, the 'decay volume'. In this volume the mesons decay to produce muon neutrinos and antineutrinos:

$$\pi^+ \to \mu^+ + \nu_\mu \qquad K^+ \to \mu^+ + \nu_\mu \qquad \text{(FHC)}$$
$$\pi^- \to \mu^- + \bar{\nu}_\mu \qquad K^- \to \mu^- + \bar{\nu}_\mu \qquad \text{(RHC)}$$

The resulting neutrino beam is dominated by $\nu_\mu/\bar{\nu}_\mu$, but there is a small intrinsic $\nu_e/\bar{\nu}_e$ contribution from decays such as

$$K^+ \to \pi^0 + e^+ + \nu_e$$

and

$$\mu^+ \to e^+ + \nu_e + \bar{\nu}_\mu$$

for example, in FHC. In the T2K beam simulation [4], 94% (82%) of ν_μ ($\bar{\nu}_\mu$) in the beam come from $\pi^\pm$ decay, and 48% (90%) of ν_e ($\bar{\nu}_e$) in the beam come from kaon decay.

In both forward and reverse horn current there will also be some amount of wrong-sign contamination ($\bar{\nu}$ in the ν beam or ν in the $\bar{\nu}$ beam) from imperfect meson focusing and meson decays which produce the wrong-sign neutrino. The wrong-sign contamination is much larger in RHC than FHC—this is demonstrated in Fig. 3.3, which shows the unoscillated neutrino flux prediction at the far detector for ν_μ, $\bar{\nu}_\mu$, ν_e, and $\bar{\nu}_e$ separately for both forward and reverse horn currents. A consequence of this larger wrong-sign contamination in RHC, combined with the fact that neutrino cross sections are around four times larger than the equivalent cross sections for antineutrino interactions (as shown in Fig. 3.16), is that the total number of right-sign interactions (ν in FHC or $\bar{\nu}$ in RHC) is much smaller in RHC than FHC.

At the end of the decay volume sits a beam dump. All hadrons from the beam, as well as muons below ~5 GeV are stopped, and any muons above ~5 GeV that pass through are measured by muon monitors [2], which sit just behind the beam dump. These measurements are used to check the stability of the neutrino beam on a bunch-

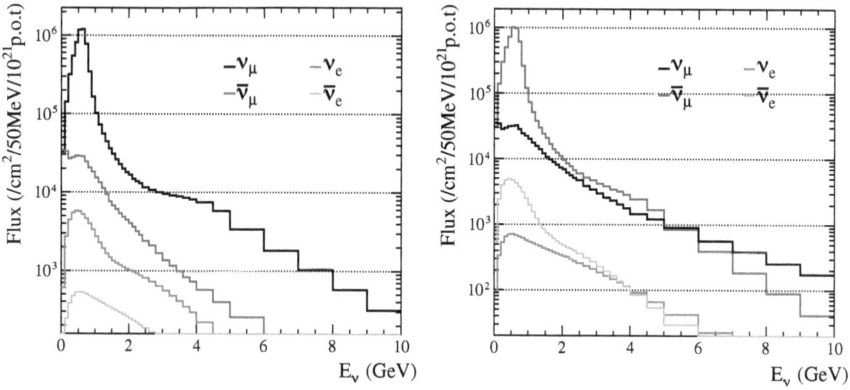

Fig. 3.3 Unoscillated neutrino flux prediction at the far detector, broken down by neutrino flavour. Copyright by the T2K Collaboration

by-bunch basis, since muons are mainly produced with neutrinos in a two-body pion decay.

Finally, the neutrinos produced from meson decay pass through the beam dump and become the T2K neutrino beam.

3.1.2 The Off-Axis 'Trick'

T2K uses an off-axis beam 'trick', in which the far detector and one of the near detectors are placed 2.5° off-axis with respect to the neutrino beam. This utilises the fact that the energy of a neutrino emitted at a large angle to the parent meson in a two-body pion or kaon decay (which is the dominant mode for neutrino production in the T2K beam) depends only weakly on the parent meson momentum.

Following the treatment in [5], consider neutrinos produced in thebreak two-body $\pi \rightarrow \mu\nu_\mu$ decay. Energy and momentum conservation lead to an expression which relates the neutrino energy E_ν to the pion energy E_π, the pion mass m_π, and the muon mass m_μ:

$$E_\nu = \frac{m_\pi^2 - m_\mu^2}{2(E_\pi - p_\pi \cos\theta)}$$

where θ is the angle relative to the pion direction at which the neutrino is emitted. Differentiating this expression with respect to E_π at constant θ, we find that E_ν has a maximum when $E_\pi = E_\pi^{max} = p_\pi / \cos\theta$. Substituting E_π^{max} into the above equation, we find the maximum neutrino energy, E_ν^{max}, for a given angle θ:

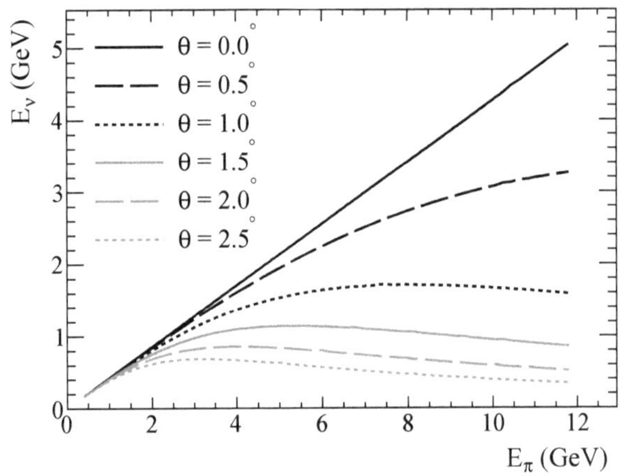

Fig. 3.4 Neutrino energy, E_ν, as a function of pion energy, E_π, for neutrinos produced in two-body $\pi \to \mu\nu_\mu$ decays. Predictions are shown for a number of off-axis angles θ between the neutrino and pion directions. T2K uses an off-axis angle of 2.5°

$$E_\nu^{max}\Big|_\theta = \frac{m_\pi^2 - m_\mu^2}{2E_\pi^{max}\sin^2\theta}$$

For pion energies both greater and less than E_π^{max}, the neutrino energy is less than E_ν^{max}. The consequence of this is that, for large emission angles θ, the range of possible neutrino energies is reduced and pions over a wide range of energies will decay to produce neutrinos with similar lab energies. This principle is demonstrated in Fig. 3.4, which shows the neutrino energy E_ν produced for a given pion energy E_π at various angles θ to the beam.

By positioning detectors at an angle to the beam we can make use of this property to effectively generate a neutrino beam with a very narrow spread of energies. Figure 3.5 shows the predicted neutrino flux on and off axis relative to the neutrino beam, demonstrating that the flux at an off-axis angle of 2.5° is much more sharply peaked as a function of neutrino energy than the on-axis flux. The beam energy and off-axis angle are chosen such that the peak neutrino energy is ~0.6 GeV, which maximises the effect of neutrino oscillation at the far detector (since the beam peak is aligned with the first oscillation maximum), and minimises backgrounds from non-oscillating neutrinos.

3.1.3 Neutrino Flux Simulation

The neutrino flux is modelled by a data-driven Monte Carlo (MC) prediction, which is tuned to in-situ measurements of the primary proton beam and magnetic horn currents, the alignment and off-axis angle of the neutrino beam, and external hadron-production measurements [4].

In the simulation, protons with a kinetic energy of 30 GeV are injected into the graphite target. The FLUKA2008 [6, 7] software is used to simulate hadronic inter-

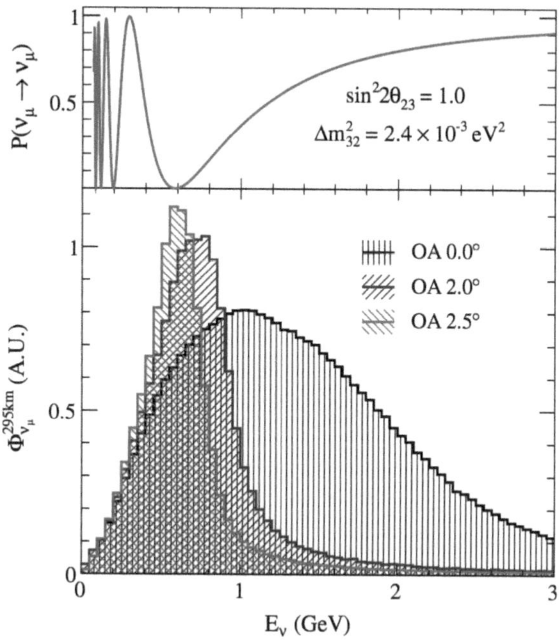

Fig. 3.5 T2K neutrino fluxes for different off-axis angles and muon neutrino survival probability at 295 km. Note that these flux predictions are normalised such that the units of the y-axis are arbitrary—in reality we also see a decrease in the total flux at higher off-axis angles. Reprinted figure with permission from K. Abe et al. (T2K Collaboration), Phys. Rev. D, vol. 87, p. 012001 (2013). Copyright (2013) by the American Physical Society [4]

actions in the target and surrounding area, where the proton beam first interacts and produces the majority of secondary particles. The FLUKA model is reweighted according to external data measurements, of which the most significant come from the NA61/SHINE experiment at CERN [8, 9]. NA61/SHINE is a dedicated hadron-production experiment at the CERN SPS which studies the particles produced when a 30 GeV proton beam interacts with a graphite target. Currently only thin (2 cm) target data are used to tune the T2K flux prediction, but data has been taken using a replica T2K target and will be incorporated into the flux prediction soon.

The particles produced in this simulation are tracked as they propagate through the secondary beamline using GEANT3 [10]. Secondary hadronic interactions are simulated with the GCALOR [11] package, with the interaction cross sections tuned to experimental data. The secondary hadrons and un-interacted protons are tracked until they either decay into neutrinos, fall below the energy threshold to decay into neutrinos, or are stopped at the beam dump. The neutrino tracks are extrapolated to the near and far detectors to produce the predicted (unoscillated) neutrino fluxes and energy spectra, which are shown in Fig. 3.3.

3.2 The T2K Near Detectors

The neutrino beam is first measured by a suite of near detectors 280 m from the proton interaction target. The effect of neutrino oscillation is expected to be negligible at this point, so the properties of the ‘unoscillated’ beam can be measured and used to tune the prediction at the far detector. Two detectors are used to measure the unoscillated neutrino beam: INGRID and ND280.

3.2.1 INGRID

The INGRID detector [1, 12] lies on axis with respect to the neutrino beam and monitors the neutrino beam direction and profile. Since the energy spectrum of the neutrino beam changes with the off-axis angle, as shown in Fig. 3.5, it is important to know the neutrino beam direction very precisely. These measurements are used indirectly in the oscillation analysis at T2K to inform the flux simulation and uncertainty.

INGRID, shown in Fig. 3.6, is made of 14 identical iron/scintillator modules arranged in a cross formation, and 2 more identical modules located at off-axis positions outside the cross. The spatial width of the neutrino beamline at INGRID is around 5 m, so the detector was designed such that the main cross spans 10 m × 10 m in the direction transverse to the beam. The centre of the cross, which has two overlapping modules, is directly aligned with the direction of the secondary beamline.

Each of these modules consists of a sandwich structure of alternating iron plates and planes of scintillator bars, which is surrounded by veto scintillator planes to

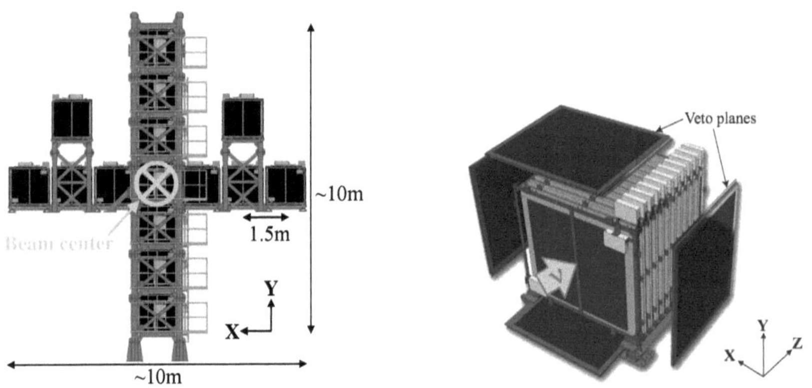

(a) INGRID. Figure from [13], copyright by the T2K Collaboration

(b) An INGRID module. The veto planes are shown in black surrounding the scintillator planes (grey) and iron plates (dark grey). Figure from [1]

Fig. 3.6 The INGRID near detector. Figure 3.6b reprinted from Nucl. Instr. Meth. Phys. Res. A, vol. 659 (1), K. Abe et al. (T2K Collaboration), “The T2K Experiment”, pp. 106–135, Copyright (2011), with permission from Elsevier [1] reject interactions taking place outside the module. Neutrino interaction events are identified by reconstructing the tracks of charged particles generated by interactions in the iron target.

There is one final module of INGRID, which is different to the other 16: the proton module. The purpose of this module is to detect both muons and recoil protons produced in charged-current quasielastic interactions at INGRID, which can be compared to MC to improve the simulation of the beamline and neutrino interactions. The proton module consists solely of scintillator planes, with no iron plates (meaning that almost all neutrino interactions in the proton module are on a carbon target, as the vast majority of nucleons in the scintillator material are in carbon nuclei), surrounded by veto planes. It is placed in the centre of the INGRID cross between the vertical and horizontal modules.

Each INGRID scintillator bar has a 3 mm-diameter hole running down the centre, which contains a wavelength-shifting (WLS) fibre to collect the light produced in the bars. The WLS fibres are each connected to a multi-pixel photon counter (MPPC) [14] at one end, which converts the light into an electronic signal. The signal is read out by a set of Trip-T front-end electronics boards (electronics boards based on the Trip-T chip, originally designed at Fermilab for the Tevatron D0 experiment [14, 15]), each of which is connected to up to 48 MPPCs via coaxial cables. The back end of the electronics system consists of readout merger modules and clock modules, which allow data to be read out from the detector, and clock and trigger signals to be sent to the front-end boards.

Sufficient statistics are collected at INGRID to allow daily measurements of the neutrino interaction event rate with 4% precision, and monthly measurements of the neutrino beam centre with accuracy better than 0.4 mrad.

3.2.2 ND280

The off-axis near detector, ND280 [1], is used directly in the oscillation analysis to reduce uncertainties in the flux and cross-section predictions. This is done by measuring the event rate of ν_μ or $\bar{\nu}_\mu$ interactions, which gives information about the unoscillated flux and energy spectrum in the direction of the far detector, as well as the rate of a number of different types of neutrino interactions. These measurements made at ND280 are used to characterise and more accurately predict the signals and backgrounds in the Super-Kamiokande detector.

ND280 is a composite detector, made up of a number of subdetectors as shown in Fig. 3.7. The most upstream subdetector is the π^0 detector, or P0D: a scintillator detector consisting of fillable water target bags and lead and brass sheets interleaved with scintillator bars. Its purpose is to measure the neutral current process $\nu_\mu + N \rightarrow \nu_\mu + \pi^0 + X$ (for some nuclei N and X) on a water target in the same neutrino flux as the far detector.

Behind the P0D is what is known as the 'tracker' region: three Time Projection Chambers (TPCs) alternating with two Fine-Grained Detectors (FGDs), described

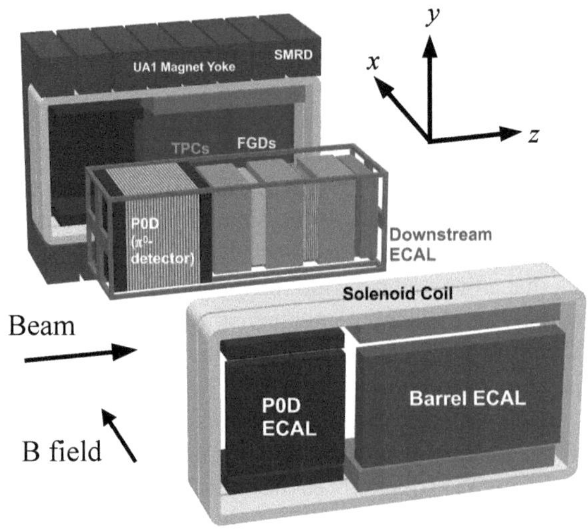

Fig. 3.7 An exploded view of the ND280 near detector. Reprinted from Nucl. Instr. Meth. Phys. Res. A, vol. 659 (1), K. Abe et al. (T2K Collaboration), "The T2K Experiment", pp.106–135, Copyright (2011), with permission from Elsevier [1]

in more detail in Sects. 3.2.2.2 and 3.2.2.3 respectively. The FGDs are designed to provide a target for neutrino interactions, and have excellent vertexing and tracking capabilities close to the interaction point. They are thin enough (~30 cm) that most of the particles produced in the interaction, especially muons, exit the FGD in which the interaction occurred and enter one of the TPCs. The TPCs are then used to measure the interaction products with very good momentum resolution and particle identification ability.

These inner detectors are encased in an electromagnetic calorimeter (ECal), which is split into two subdetectors: the P0D ECal (which surrounds the P0D) and the tracker ECal (which surrounds the tracker region). Both are made of layers of plastic scintillator alternating with lead absorber sheets. The ECals provide near-hermetic coverage for all particles exiting the inner detector volume, and are used in full event reconstruction to detect photons and measure their energy and direction, as well as providing extra information about high-energy charged particles.

Finally, the entire ND280 detector sits in a 0.2 T magnetic field generated by the repurposed UA1 magnet from CERN,[1] described in more detail in Sect. 3.2.2.1. The magnet is instrumented with scintillator to act as a side muon ranging detector (SMRD). This has multiple functions: to measure muons escaping the detector at high angles relative to the beam direction; to form part of the trigger for cosmic ray muons that enter the ND280 detector; and to identify beam-related interactions in the iron of the magnet and the surrounding cavity.

The outer dimensions of the UA1 magnet are 7.6 m × 5.6 m × 6.1 m (length z × width x × height y), and the basket which contains the inner subdetectors (P0D, TPCs, and FGDs) measures 6.5 m × 2.6 m × 2.4 m.

[1] The magnet was also used by the NOMAD experiment before being shipped to Japan for T2K.

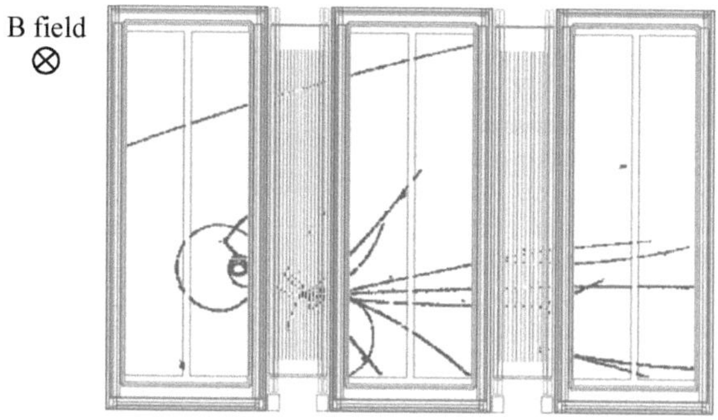

Fig. 3.8 An event display in ND280, shown for the tracker detectors (FGDs and TPCs) only. Reprinted from Nucl. Instr. Meth. Phys. Res. A, vol. 637 (1), N. Abgrall et al. "Time Projection Chambers for the T2K Near Detectors", pp. 25–46, Copyright (2011), with permission from Elsevier [16]

The oscillation analysis described in this thesis mostly uses information from the tracker region (the TPCs and FGDs), which relies on the magnetic field produced by the UA1 magnet. Therefore only these subdetectors will be described in more detail. Further details about the other subdetectors in ND280 can be found in [1] and the references therein.

Figure 3.8 shows the tracker-only portion of an event display from ND280. Two neutrino interactions are shown in the same event: one neutrino interacted in front of the first TPC, producing a single charged particle at the top-left of the figure, and a second neutrino undergoes a deep inelastic scattering interaction in the first FGD, producing a number of charged particles which are measured in all three TPCs and both FGDs.

3.2.2.1 UA1 Magnet

The repurposed UA1 magnet [1] provides a dipole magnetic field of 0.2 T in ND280, which allows the TPCs to measure the momentum of charged particles with resolution better than 10% ($\delta p_{\perp}/p_{\perp} < 0.1 p_{\perp}$, where $p_{\perp}$ is the momentum perpendicular to the magnetic field dierection), and to determine the sign of charged particles (thus identifying whether a CCQE interaction that produced a single charged particle was instigated by a neutrino or an antineutrino). The magnet consists of aluminum coils, which create the dipole field, and a flux return yoke. The coils are water-cooled: they are made from aluminum bars with a central bore that allows water to flow inside the bars and cool the magnet.

The magnetic field was measured in-situ before the start of data-taking, with an uncertainty of 2×10^{-4} T for each field component at the nominal field of 0.2 T. During data-taking the current is monitored and used to calculate the actual magnetic field, which helps to reduce the systematic uncertainty of the momentum determination in the TPCs.

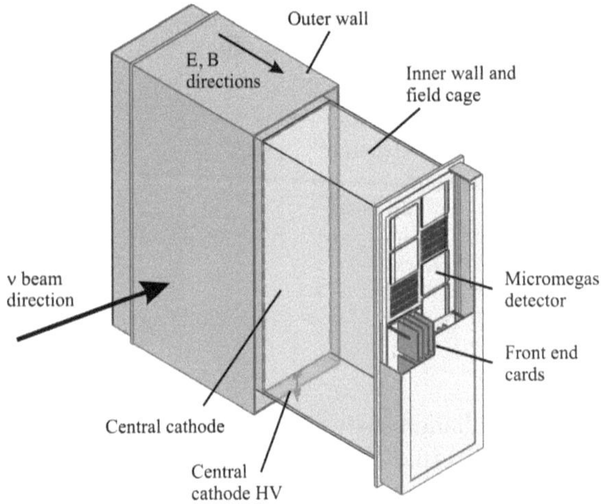

Fig. 3.9 A cut-away drawing showing the main aspects of the TPC design. The outer dimensions of the TPC are approximately 2.3 m × 2.4 m × 1.0 m. Reprinted from Nucl. Instr. Meth. Phys. Res. A, vol. 637 (1), N. Abgrall et al. "Time Projection Chambers for the T2K Near Detectors", pp. 25–46, Copyright (2011), with permission from Elsevier [16]

3.2.2.2 Time Projection Chambers

The three TPCs [16] all have identical designs and alternate with the two FGDs in the ND280 tracker. Because they are able to image events in three dimensions, the TPCs can easily determine the number and orientations of charged particle tracks. This can be used to select and study high-purity samples of different types of neutrino interaction in ND280. The magnetic field provided by the UA1 magnet allows the TPC information to be used to measure the momenta of charged particles and therefore determine the event rate at ND280 as a function of reconstructed (neutrino) energy before oscillation.

Each TPC module consists of two boxes, one inside the other, as shown in Fig. 3.9. The inner box holds an argon-based drift gas, and the walls of the inner box form the field cage. The walls of the outer box are held at ground potential, and the outer box is filled with CO_2 gas to provide electrical insulation between the two boxes. The inner box is subdivided by a central cathode at its midpoint, and the walls parallel to the cathode are covered with panels that were precisely machined into a copper strip pattern. This produces a uniform electric drift field inside the module, aligned with the magnetic field produced by the magnet.

Charged particles passing through a TPC module produce ionization electrons in the gas. These drift away from the cathode towards the readout planes on the walls of the inner box, where the electrons are multiplied and detected using bulk micromegas detectors [17, 18]. The arrival times and positions of signals in the readout plane are combined to give complete 3D images of the particle paths.

Twelve micromegas modules, segmented into 7.0 mm × 9.8 mm pixels, cover each readout plane in two offset columns, so that the inactive regions are not aligned. In total across all three TPCs there are 72 micromegas modules, with almost 9 m^2 of active surface. The active tracking length of each TPC module is approximately 720 mm, and the maximum drift distance from the central cathode to the micromegas

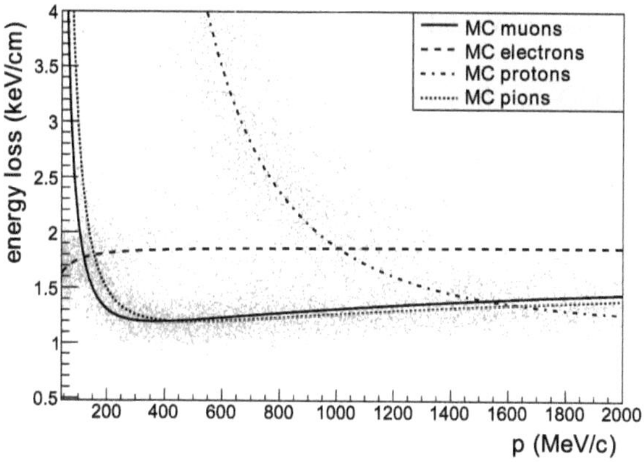

Fig. 3.10 The distribution of dE/dx (labelled 'energy loss') in one of the TPCs as a function of momentum for positively charged particles. The points show measurements by a single TPC of the energy loss and momentum of positively charged particles produced in neutrino interactions, and the curves show the simulated expectations. Reprinted from Nucl. Instr. Meth. Phys. Res. A, vol. 659 (1), K. Abe et al. (T2K Collaboration), "The T2K Experiment", pp. 106–135, Copyright (2011), with permission from Elsevier [1]

detectors is 897 mm. The TPC modules are constructed entirely out of non-magnetic materials, so as not to distort the field provided by the magnet. The amount of material in each TPC is kept as small as possible, particularly in the walls between the tracking volume and the FGDs, where the charged particles used to infer the neutrino energy spectrum could scatter.

It is possible to distinguish between different types of charged particles in the TPCs using the energy loss per distance, dE/dx. This is demonstrated in Fig. 3.10, which shows a scatter plot of dE/dx (labelled 'energy loss') by positively charged particles in the TPCs as a function of momentum. The solid, dashed, dot-dashed, and dotted lines show the expected results for muons, electrons, protons, and pions respectively. The resolution on dE/dx is $7.8 \pm 0.2\%$ for minimum-ionizing particles, better than the design requirement of 10%, which allows muons to be successfully distinguished from electrons in the TPCs.

3.2.2.3 Fine-Grained Detectors

The purpose of the two FGDs [19] is to provide a target mass for neutrino interactions in ND280 and to track charged particles close to the interaction vertex. The TPCs provide excellent 3D tracking and particle identification for charged particles which exit the FGDs, but it is also important to accurately measure short-ranged particles such as recoil protons, in order to identify the type of neutrino interaction that occurred. The FGDs must therefore be able to measure charged particles with fine granularity so that individual particle tracks can be resolved and their directions measured.

There are two FGDs, each of which contains 1.1 tons of target material (this mass is required to yield a sufficient statistical sample of neutrino interactions). Both FGDs contain polystyrene scintillator bars, 1864.3 mm long and oriented perpendicular to the beam in either the x or y direction. To achieve the required granularity, each bar has a square cross section of 9.6 mm on each side. The scintillator bars are each covered with a reflective coating containing TiO_2, and have a wavelength shifting (WLS) fibre running down the centre. Each WLS fibre is read out at one end by a multi-pixel photon counter (MPPC, chosen instead of photomultiplier tubes because it must operate in a magnetic field), with the other end mirrored by vacuum deposition of aluminium.

The scintillator bars are arranged into 'XY modules', with each module consisting of 192 scintillator bars in the horizontal direction and 192 scintillator bars in the vertical direction. FGD1 is made purely from these scintillator bars, arranged into 15 XY modules, which provide the entire target mass for neutrino interactions. FGD2 contains 7 XY scintillator modules alternating with 6 2.5 cm thick layers of water. This makes it possible to measure neutrino cross sections on water at ND280, which is vital for accurately predicting the number of expected events at the far detector. All relevant neutrino cross sections depend at some level on the target nucleus, and nuclear effects cannot be reliably corrected from one material to another by theoretical calculations. Therefore it is useful to be able to measure interaction rates on water at the near detector in order to predict the rates of these processes at Super-Kamiokande.

The main aim of particle identification in the FGDs is to distinguish stopping protons from stopping pions (protons and pions that don't exit the FGD). The TPCs have much better particle identification capability than the FGDs, so any particles which do exit the FGDs into the TPCs are measured there. Particle identification in the FGDs is done using the same technique as in the TPCs: the energy loss per track length, $\mathrm{d}E/\mathrm{d}x$. However, the resolution and track length in the FGDs is too low to allow a calculation of $\mathrm{d}E/\mathrm{d}x$ for measured tracks in data. Instead, the prediction is integrated to find the expected range of a particle with a given energy—low energy particles with the same kinetic energy but different masses can have significant differences in track lengths. The track length is calculated as a straight line between the initial and final 3D reconstructed positions of the particle in the FGD, and the measured light yield for hits along a track is converted into an equivalent energy deposit, which is summed over all hits in a track. The charged particle can then be identified by comparing the total energy deposited as a function of total track length to the theoretical expectation. This is demonstrated in Fig. 3.11, which shows a scatter plot of total deposited energy as a function of particle range for charged particles produced by neutrino interactions in FGD1. The solid, dashed, and dot-dashed lines show the expected results for protons, muons, and pions respectively, and demonstrate how this technique can be used to distinguish protons from other charged particles.

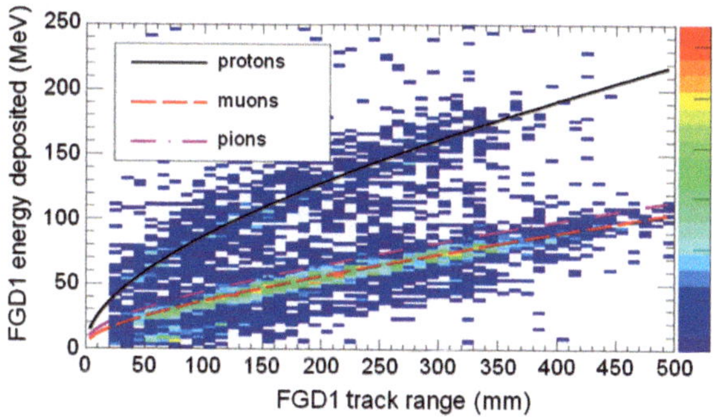

Fig. 3.11 The distribution of total energy deposited in FGD1 as a function of track range for particles stopping in FGD1. The coloured 2D plot shows the number of events per bin from stopping particles in neutrino beam data (using a linear scale with arbitrary units), and the curves show the MC expectations. Reprinted from Nucl. Instr. Meth. Phys. Res. A, vol. 696, P.-A. Amaudruz et al. "The T2K Fine-Grained Detectors", pp. 1–31, Copyright (2012), with permission from Elsevier [19]

3.2.3 Near Detector Simulation

Neutrino interactions in both detectors are simulated using the NEUT generator [20], as described in Sect. 3.4.1. The final-state particles are propagated through the detectors using GEANT4 [10], which includes a detailed simulation of the detector geometry, for both ND280 and INGRID, and the experimental hall. The scintillator, WLS fibre, MPPC, and electronics response is simulated with a custom-written software package called elecSim, which uses information from empirical analytical models for propagation of light in the fibres.

3.3 The T2K Far Detector: Super-Kamiokande

The far detector, Super-Kamiokande (SK) [21], shown in Fig. 3.12, is a 50 kton (22.5 kton fiducial mass) water Cherenkov detector. It is located 295 km west of the near detectors, 1 km deep within the centre of Mt. Ikenoyama, and is filled with pure water.

SK consists of a cylindrical tank, 39 m in diameter and 42 m tall. The tank is separated into two major volumes, an inner and an outer detector, by a cylindrical stainless-steel framework around 55 cm thick and placed approximately 2–2.5 m inside the tank walls on all sides. This steel structure supports both inward-facing and outward-facing PMTs. The outer detector is used primarily as an active veto of cosmic ray muons and other backgrounds, and is capable of almost 100% efficiency rejecting cosmic ray backgrounds. It also shields the inner detector from interactions in the surrounding rock. The inner detector is instrumented with 11,129 inward-facing 50 cm-diameter PMTs, giving around 40% photo-multiplier tube (PMT) coverage on

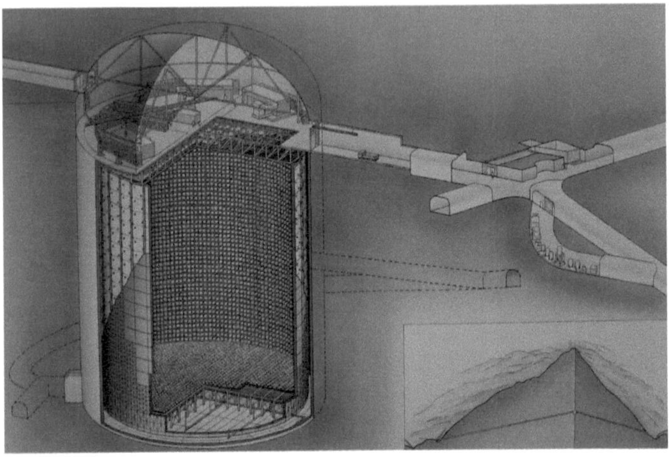

Fig. 3.12 The Super-Kamiokande detector.Figure from [**? ? ?**]

its surrounding walls. This provides sufficient spatial resolution to reconstruct the positions of products of neutrino interactions, and measurements taken in the inner detector are used directly in the T2K oscillation analyses.

Neutrino interactions in SK often produce charged particles which—if above an energy threshold—produce Cherenkov light in a cone as they travel through the water. The photons from this Cherenkov light are detected by the PMTs on the walls of the detector, where they form a ring-shaped hit pattern. The hit pattern, combined with timing information, makes it possible to extract information about the event vertex position, particle momentum, and the flavour of charged leptons.

The T2K oscillation analysis strategy relies on identifying CCQE interactions, in which only one particle will be detected in SK: a muon from a ν_μ interaction, or an electron from a ν_e interaction. If the interaction is truly CCQE, the neutrino energy can be reconstructed from the momentum and direction measurements of the outgoing lepton (to an accuracy limited by the unknown initial energy and momentum of the nucleon state with which the neutrino interacts, as discussed in Sect. 3.4) and the neutrino flavour can be inferred by identifying the flavour of the lepton produced.

Muons have a relatively large mass and are therefore fairly immune to changes in their momentum due to scattering on particles in the water. This means that muons will produce a sharp, clear Cherenkov ring. Electrons, on the other hand, with their much lighter mass, are prone to scattering in the water, and also almost always induce electromagnetic showers as they travel. Both of these effects result in a "fuzzy" ring pattern seen by the PMTs on the detector walls (which can actually be better thought of as a superposition of multiple overlapping Cherenkov rings).

Figure 3.13 shows example Cherenkov rings from a muon and an electron event in SK data. This difference between sharp and fuzzy rings can be used to discriminate between muons and electrons with very good efficiency—the particle identification algorithm was verified using a test beam experiment at KEK [23] and it is estimated

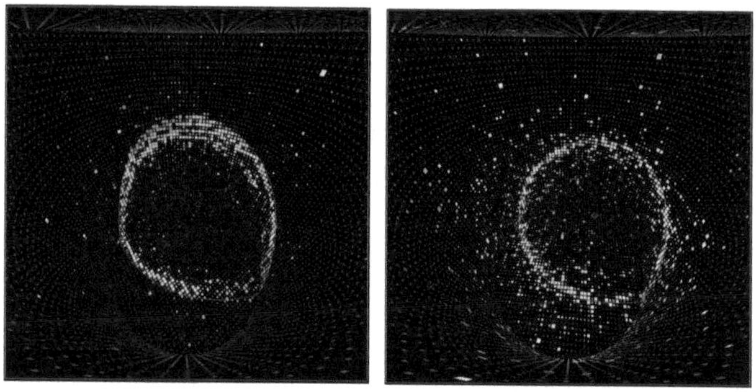

(a) Muon event (b) Electron event

Fig. 3.13 Cherenkov rings from a muon and an electron event in SK data. The size of the square representing each PMT corresponds to the amount of light observed. Figure from [25]. Copyright Kamioka Observatory, ICRR (Institute for Cosmic Ray Research), The University of Tokyo

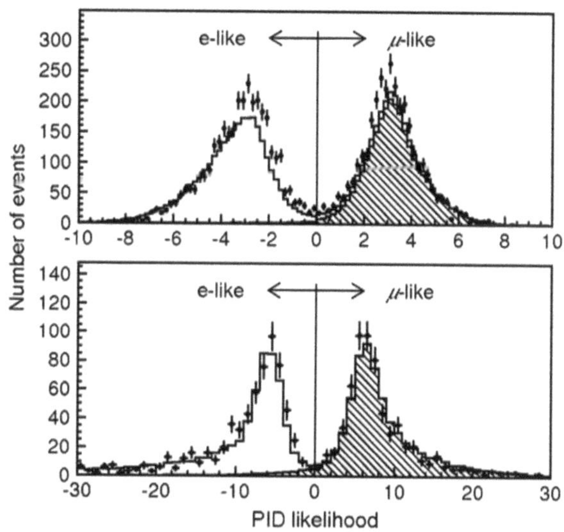

Fig. 3.14 The distribution of the particle identification likelihood for sub-GeV (*top*) and multi-GeV (*bottom*) fully-contained single-ring events in the SK detector. Both data (points) and atmospheric neutrino Monte Carlo predictions (histograms) are shown. Reprinted figure with permission from Y. Ashie et al. (Super-Kamiokande Collaboration), Phys. Rev. D, vol. 71, p. 112005 (2005). Copyright (2005) by the American Physical Society [24]

(using simulated CCQE events) that the probability for a muon event to be misidentified as an electron is 0.7% [24]. The distribution of the likelihood variable used to discriminate single-ring electron and muon events is shown in Fig. 3.14 for both data and Monte Carlo prediction, at sub-GeV and multi-GeV energies. The Monte Carlo replicates the data well and clear separation is seen in both.

An important consideration for the inclusion of antineutrino data in the oscillation analysis is that the Super-Kamiokande detector does not have a magnetic field, so is unable to differentiate between positive and negative leptons (produced by antineutrino and neutrino interactions respectively). Because of this, the neutrino and antineutrino component of the beam are measured at the ND280 near detector in order to accurately predict the observables at SK.

SK has been running since 1996, and the long-running operation of the detector means that its behaviour is well understood. Cosmic-ray muon and atmospheric neutrino interactions provide control samples of data events not related to the T2K beam. These show good agreement between data and simulated samples, and are used to assess the detector response and systematic uncertainty. The uncertainty on the calibration of the energy scale, for example, is estimated by comparing data measurements and Monte Carlo simulations of the reconstructed momentum spectrum of electrons produced in the decay of cosmic ray muons, the reconstructed mass of neutral pions from atmospheric neutrino interactions, and measurements of cosmic ray muons that stop within the SK tank. This results in a total estimated uncertainty of 2.4%.

3.3.1 *Super-Kamiokande Simulation*

Neutrino interactions in SK are modelled using NEUT, as described in Sect. 3.4.1. The propagation of the resulting particles in the Super-Kamiokande detector is simulated by a package called SKDETSIM, which is based on GEANT3. The CALOR [11] physics package is used for hadronic interactions in water, and custom routines have been written to model pions with momentum below 500 MeV [26].

3.4 Neutrino Interactions at T2K

Neutrino oscillation measurements rely on examining the composition of a neutrino beam and how it changes as a function of energy and distance travelled, which, in practice, often comes down to measuring the number of neutrino events of a given flavour and comparing the measurement to a prediction. In T2K, this is done twice: once at ND280, close to the neutrino source, to measure the 'unoscillated' spectrum, and once at Super-Kamiokande, after the neutrinos have oscillated. The measurement at the far detector is compared to the unoscillated measurement from the near detector (extrapolated forwards into a full far-detector prediction) and any deficit or excess of neutrinos interpreted as evidence of neutrino oscillation. Disregarding detector effects, the number of events measured at each detector will be a convolution of the neutrino flux (or how many neutrinos are actually there) and the neutrino interaction cross section (or how likely it is that a neutrino will interact given that it is there). Therefore, it is important to have a good understanding of neutrino interaction cross sections in order to accurately predict the observables at the far detector, as any uncertainty or lack of understanding will form an important systematic error when comparing measurements and predictions. An overview of the theory and current knowledge of neutrino interactions is given in [27], and general particle physics

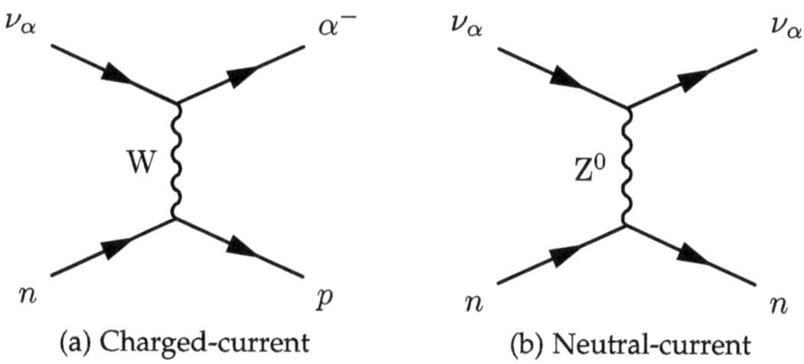

Fig. 3.15 Example tree-level Feynman diagrams for charged-and neutral-current scattering of a neutrino of flavour α from a neutron

textbooks such as [28] provide a useful introduction to weak interactions in the Standard Model.

We define two broad categories of neutrino interaction in the Standard Model: charged-current (CC) and neutral-current (NC) weak interactions. Example Feynman diagrams for each type of interaction are shown in Fig. 3.15. Only charged-current interactions are useful for the T2K oscillation measurements presented in this thesis because they contain information about the neutrino flavour (the neutrino flavour must match the flavour of the charged lepton produced in the interaction), and therefore allow us to determine whether neutrinos have changed flavour between the near and far detectors. Neutral-current interactions produce charged leptons whose flavour is uncorrelated with the interacting neutrino (for example, a ν_μ, ν_e, or ν_τ can produce an electron via neutral-current scattering). This forms an irreducible background to the selected events at the far detector—it is not possible to distinguish an electron produced in a CC interaction from one produced in an NC interaction. Instead, the NC interaction rate is estimated using Monte Carlo and included in the predictions at Super-Kamiokande.

Figure 3.16 shows the charged-current cross section measurements and predictions for neutrino and antineutrino scattering as a function of neutrino or antineutrino energy. Three different charged-current interaction modes are relevant at the energies shown: quasi-elastic scattering (QE), resonance production (RES), and deep inelastic scattering (DIS). At the neutrino energies used in T2K (the neutrino energy spectrum peaks around 0.6 GeV), most interactions involve a neutrino scattering from an entire nucleon, as shown in Fig. 3.15, rather than its constituent particles. This is known as "quasi-elastic" scattering because the 2-particle $\rightarrow$ 2-particle kinematics resemble those of an elastic scattering interaction. Additional contributions to the neutrino and antineutrino scattering cross sections come from from RES and DIS interactions in the high-energy tail of the neutrino beam.

Feynman diagrams showing the charged-current quasi-elastic (CCQE) interaction channels for neutrinos and antineutrinos are given in Fig. 3.17. The proton produced in CCQE scattering in Super-Kamiokande is usually free (i.e. it escapes the nucleus) but below Cherenkov threshold, so is not observed in the detector. However, because

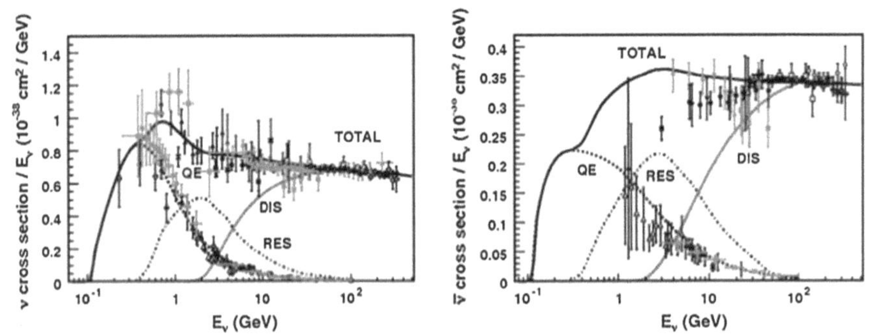

(a) ν_μ interactions (b) $\bar{\nu}_\mu$ interactions

Fig. 3.16 Muon neutrino and antineutrino charged-current cross-section measurements and predictions as a function of neutrino or antineutrino energy. Data are shown from a number of different measurements, represented by different colours and marker symbols. For full details of the data included, see [27]. The total prediction according to the NUANCE generator [29] is shown, as well as the prediction for various contributing processes separately. The dominant contributing process at the T2K peak energy is quasi-elastic (QE, *dashed*) scattering, with contributions from resonance production (RES, *dot-dashed*) and deep inelastic scattering (DIS, *dotted*) in the high-energy tail of the neutrino spectrum. Note the different scales on the y-axes: the cross sections for neutrinos are around four times larger than the equivalent cross sections for antineutrinos. Reprinted figure with permission from J. A. Formaggio and G. P. Zeller, Rev. Mod. Phys., vol. 84, pp. 1307–1341 (2012). Copyright (2012) by the American Physical Society [27]

it is a two-body process it is possible to reconstruct the neutrino energy from only the energy and direction of the final lepton, with an accuracy limited by how well the initial nucleon energy and momentum can be known. Assuming that the interaction occurs on a bound neutron at rest, we can reconstruct the neutrino energy in a ν_μ CCQE interaction:

$$E_{rec} = \frac{m_p^2 - (m_n - E_b)^2 + m_\mu^2 + 2(m_n - E_b)E_\mu}{2(m_n - E_b - E_\mu + p_\mu \cos\theta_\mu)} \tag{3.1}$$

where E_{rec} is the reconstructed neutrino energy; m_p, m_n, and m_μ are the proton, neutron and muon masses respecvively; E_b is the binding energy of the neutron; E_μ is the muon energy; p_μ is the muon momentum; and $\cos\theta_\mu$ is the angle of the muon with respect to the incoming neutrino direction. Of course, in reality nucleons are not at rest in a nucleus; uncertainties in the nucleon momentum, as well as the binding energy, will form systematic uncertainties in the reconstructed neutrino energy.

In addition to having a relatively simple relationship between the lepton kinematics and neutrino energy, the CCQE interaction cross section is also relatively well-modelled theoretically, and well constrained by data. For these reasons, CCQE interactions are used to search for neutrino oscillation at SK.

Although CCQE is the dominant interaction mode for T2K, it is important to consider backgrounds due to other interaction processes. These can be particularly

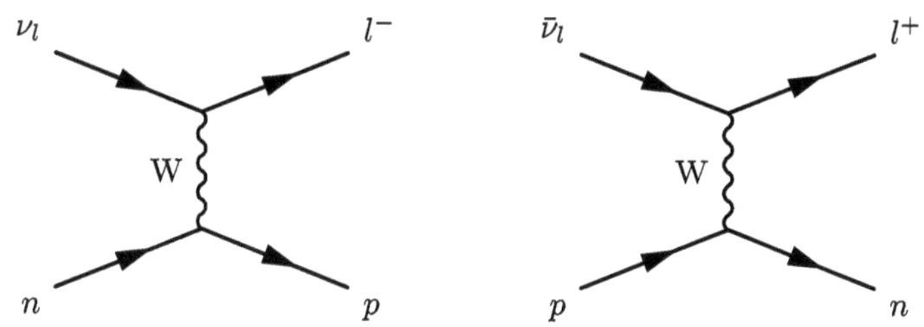

(a) ν_l CCQE scattering

(b) $\bar{\nu}_l$ CCQE scattering

Fig. 3.17 Feynman diagrams for CCQE scattering of neutrinos and antineutrinos of flavour l

problematic if the process is misidentified: interaction modes are categorised in data by the configuration of charged particles measured in the detector. If particles are missed or misidentified, this could cause an interaction to mistakenly be categorised as CCQE, leading to the neutrino energy being misreconstructed.

An important example of this at T2K is CC single pion (CC1π) production, which often happens in CC RES interactions: $\nu_l + N \rightarrow l^- + N' + \pi$. Because this interaction has a three-body final state, the initial neutrino energy is not a simple function of the lepton momentum and direction, as in Eq. 3.1. These events are excluded from the analysis at the far detector by selecting only events in which a single Cherenkov ring is observed in the detector, implying that there was only a single charged lepton in the final state. Data samples selected in this way are referred to as 'single-ring', or '1R' event samples. However, for events in which—in addition to the charged lepton—a pion is emitted below Cherenkov threshold, only a single Cherenkov ring will be observed in the detector. These events form an irreducible background to CCQE interactions in the single-ring selections at the far detector.

An additional background to the single-ring electron-like (1R$_e$) sample in Super-Kamiokande comes from neutral current interactions in which a π^0 is produced. The π^0 will decay in the detector to produce two photons, which Compton scatter and produce e^+–e^- pairs. These result in Cherenkov rings that are indistinguishable from those produced by electrons from CCQE scattering. If one of the two rings is not reconstructed or the rings overlap, it will look identical to a ν_e CCQE event.

Further complications come from final state and secondary interactions (FSI and SI) in which the particles produced in the interaction re-interact either in the nucleus or in the detector before they are detected, causing a disconnect between the interaction products and that particles which are measured. For example, the charged pion produced in a CC RES event could be absorbed in the nucleus and not escape. In this case, there would be no pion measured in the detector, and we might classify the event as CCQE. However, because the event was not truly CCQE the neutrino energy would not correctly be described by the CCQE formula given in Eq. 3.1.

The interactions mentioned above, as well as CCQE and other non-CCQE interaction modes, are measured at ND280 to improve the prediction at Super-Kamiokande and estimate the size of the background to the CCQE sample due to non-CCQE interactions.

3.4.1 Neutrino Interaction Simulation

Interactions of neutrinos inside the detectors of T2K are simulated by default using NEUT version 5.3.2. The NEUT neutrino interaction generator is described in detail in [20], and a thorough discussion of the interaction models used for the analyses described in this thesis is given in [30–33].

The default NEUT model for CCQE interactions is the relativistic Fermi gas (RFG) model by Smith and Moniz [34], which treats all bound nucleons as quasi-free, and has historically been used in many neutrino interaction generators because of its simplicity. However, the spectral function (SF) model by O. Benhar and collaborators [35] has recently been implemented, which gives a more realistic description of the momentum and energy distributions of nucleons within a nucleus than the RFG model. This model uses the impulse approximation: it assumes that the neutrino interacts with one nucleon only but allows for short-range correlations between nucleons, which affect the energy and momentum distribution for the nucleon involved in the interaction.

Another model that has been recently added to NEUT is the random phase approximation (RPA) model [36], which parameterises a correction to the W propagator due to nuclear screening effects arising from long-range inter-nucleon correlations (as opposed to the short-range correlations considered in the SF model). The calculation itself is based on a local Fermi gas (LFG) model of the nucleus, which is not currently available in NEUT. However, the authors have said that it can be applied with reasonable precision to the RFG model in NEUT. Two RPA calculations are available from the same authors, which we label 'relativistic' and 'non-relativistic'.

The most significant update to NEUT is the inclusion of a multi-nucleon neutrino interaction model from Nieves et al. [36, 37] (referred to as either '2p–2h' or 'MEC', for 'meson exchange currents', in this thesis). This goes beyond the impulse approximation described in the SF model because the interaction is not with a single quasi-free nucleon, but may occur on a multi-nucleon state. Although these interactions are not truly CCQE, they are indistinguishable from true CCQE interactions in SK, and significantly enhance the CCQE-like neutrino interaction cross section.

The new default T2K NEUT simulation uses the SF+MEC model for CCQE interactions. However, the *Neutrino Interactions Working Group* (NIWG) in T2K found that the RFG+rel.RPA+MEC model fit better to both T2K and external CCQE data [30], so the ND280 and SK MC is reweighted to be consistent with the predictions of this model for all MC predictions presented in this thesis.

The CC RES pion production model in NEUT is from Rein and Sehgal [38], but retuned using form factors from Graczyk and Sobczyk [39], with parameter values determined by a reanalysis of the ANL and BNL bubble chamber data sets [40].

3.5 Data Analysed

T2K has been taking data since 2010, and since mid-2014 the beam has been run in mostly antineutrino mode.

Table 3.1 shows the data collected by T2K (in units of 10^{20} protons on target, or POT), separated into 'runs' based on when the data were collected. Runs 1–4 used entirely neutrino-mode beam, and runs 5–7 used mostly antineutrino-mode beam, with a small amount of neutrino-mode running. The analysis presented in this thesis uses data from ND280 and SK, and the amount of data analysed from each detector is outlined in the table.

In general, a large proportion of the data provided is analysed for SK, with some data loss due to data quality requirements. A smaller proportion of the provided data are analysed for ND280, partially because the more complicated detector means that a larger proportion of the data fail the data quality cuts, and also because some T2K runs have been excluded from the analysis. From December 2011 to March 2012, the T2K beam was run with the magnetic horns turned off, and this data set is not included at either detector. In T2K run 1 (January to June 2010), the ND280 detector was not fully operational so this data set is also excluded from the ND280 analysis. The neutrino-mode run 5, 6, and 7 data and antineutrino-mode run 7 data were not included at ND280 (or the neutrino-mode run 7 data at SK) because of time constraints—it was not possible to calibrate and prepare the data in time for the analysis. Although the amount of neutrino-mode data not analysed at ND280 for this

Table 3.1 The data, in units of 10^{20} protons on target (POT), which was provided at T2K and analysed in this thesis

T2K Run	Dates	Data provided	Data analysed	
			ND280	SK
1 (ν mode)	Jan. 2010–Jun. 2010	0.33	0.00	0.32
2 (ν mode)	Nov. 2010–Mar. 2011	1.12	0.78	1.11
3 (no horns)	Dec. 2011–Mar. 2012	0.01	0.00	0.00
3 (ν mode)	Mar. 2012–Jun. 2012	1.59	1.56	1.58
4 (ν mode)	Oct. 2012–May 2013	3.60	3.47	3.56
5 (ν mode)	May 2014–Jun. 2014	0.24	0.00	0.24
5 ($\bar{\nu}$ mode)		0.51	0.43	0.51
6 (ν mode)	Nov. 2014–Jun. 2015	0.21	0.00	0.19
6 ($\bar{\nu}$ mode)		3.53	2.41	3.51
7 (ν mode)	Feb. 2016–May 2016	0.09	0.00	0.00
7 ($\bar{\nu}$ mode)		3.49	0.00	3.46
Total ν mode		7.19	5.82	7.00
Total $\bar{\nu}$ mode		7.53	2.84	7.47

reason is fairly small, a larger amount of antineutrino-mode data has been excluded, and will be included in future analyses.

The analysis presented in this thesis uses data from T2K runs 1–7b, which were collected between January 2010 and May 2016. This amounts to a total of 7.00×10^{20} POT in neutrino mode and 7.47×10^{20} POT in antineutrino mode at SK, and 5.82×10^{20} POT in neutrino mode and 2.84×10^{20} POT in antineutrino mode at ND280. Two other analyses are described in Appendix A, which use the same neutrino-mode data set from ND280 but only the antineutrino-mode data from run 5 at ND280 and runs 5 and 6 at SK, giving a total of 0.43×10^{20} POT in antineutrino mode at ND280 and 4.01×10^{20} POT in antineutrino mode at SK.

References

1. K. Abe et al. (T2K Collaboration), The T2K experiment. Nucl. Inst. Methods Phys. Res. Sect. A Accel. Spectr. Detec. Assoc. Equ. **659**(1), 106–135 (2011). doi:10.1016/j.nima.2011.06.067
2. K. Suzuki et al., Measurement of the muon beam direction and muon flux for the T2K neutrino experiment. Progr. Theor. Exp. Phys. **2015**(5) (2015). doi:10.1093/ptep/ptv054
3. T. Sekiguchi et al., Development and operational experience of magnetic horn system for T2K experiment. Nucl. Instr. Methods Phys. Res. Sect. A Accel. Spectr. Detec. Assoc. Equ. **789**, 57–80 (2015). doi:10.1016/j.nima.2015.04.008
4. K. Abe et al. (T2K Collaboration), T2K neutrino flux prediction. Phys. Rev. D **87**, 012001 (2013). doi:10.1103/PhysRevD.87.012001
5. D. Beavis et al., Long baseline neutrino oscillation experiment, E889, Physics Design Report, Chapt. III A, BNL 52459 (1995), http://puhep1.princeton.edu/mcdonald/nufact/e889/chapter3a.pdf
6. G. Battistoni et al., The FLUKA code: description and benchmarking. AIP Conf. Proc. **896**(1), 31–49 (2007). doi:10.1063/1.2720455
7. A. Ferrari, P.R. Sala, A. Fasso, J. Ranft, FLUKA: a multi-particle transport code, Report No. CERN-2005-010; Report No. SLAC-R-773; Report No. INFN-TC-05-11 (2005), http://slac.stanford.edu/pubs/slacreports/reports16/slac-r-773.pdf
8. N. Abgrall et al. (NA61/SHINE Collaboration), Measurements of cross sections and charged pion spectra in proton-carbon interactions at 31 GeV/*c*. Phys. Rev. C **84**, 034604 (2011). doi:10.1103/PhysRevC.84.034604
9. N. Abgrall et al. (NA61/SHINE Collaboration), Measurement of production properties of positively charged kaons in proton-carbon interactions at 31 GeV/*c*. Phys. Rev. C **85**, 035210 (2012). doi:10.1103/PhysRevC.85.035210
10. R. Brun, F. Carminati, S. Giani, GEANT: detector description and simulation tool, Technical Report No. CERN-W5013 (1994), http://www.slac.stanford.edu/comp/physics/cernlib/cerndoc/geant/aaaa.ps.gz
11. C. Zeitnitz, T.A. Gabriel, The GEANT-CALOR interface and benchmark calculations of ZEUS test calorimeters. Nucl. Inst. Methods Phys. Res. Sect. A Accel. Spectr. Detec. Assoc. Equ. **349**(1), 106–111 (1994). doi:10.1016/0168-9002(94)90613-0
12. K. Abe et al., Measurements of the T2K neutrino beam properties using the INGRID on-axis near detector. Nucl. Inst. Methods Phys. Res. Sect. A Accel. Spectr. Detec. Assoc. Equ. **694**, 211–223 (2012). doi:10.1016/j.nima.2012.03.023
13. T. Kikawa et al., Flux prediction and uncertainty updates with NA61 2009 thin target data and negative focussing mode predictions, Technical report 160, T2K (2014), http://www.t2k.org/docs/technotes/160

14. A. Vacheret et al., The front end readout system for the T2K-ND280 detectors, in *2007 IEEE Nuclear Science Symposium Conference Record*, vol. 3 (2007), pp. 1984–1991. doi:10.1109/NSSMIC.2007.4436543
15. J. Estrada, C. Garcia, B. Hoeneisen, P. Rubinov, MCM II and the trip chip, Technical report 2226, FERMILAB (2002), https://inspirehep.net/record/635982/files/fermilab-tm-2226.PDF
16. N. Abgrall et al., Time projection chambers for the T2K near detectors. Nucl. Inst. Methods Phys. Res. Sect. A Accel. Spectr. Detec. Assoc. Equ. **637**(1), 25–46 (2011). doi:10.1016/j.nima.2011.02.036
17. Y. Giomataris, J.P. Ph Rebourgeard, G.Charpak Robert, MICROMEGAS: a high-granularity position-sensitive gaseous detector for high particle-flux environments. Nucl. Inst. Methods Phys. Res. Sect. A Accel. Spectr. Detec. Assoc. Equ. **376**(1), 29–35 (1996). doi:10.1016/0168-9002(96)00175-1
18. I. Giomataris et al., Micromegas in a bulk. Nucl. Inst. Methods Phys. Res. Sect. A Accel. Spectr. Detec. Assoc. Equ. **560**(2), 405–408 (2006). doi:10.1016/j.nima.2005.12.222
19. P.-A. Amaudruz et al., The T2K fine-grained detectors. Nucl. Instr. Methods Phys. Res. Sect. A Accel. Spectr. Detec. Assoc. Equ. **696**, 1–31 (2012). doi:10.1016/j.nima.2012.08.020
20. Y. Hayato, A neutrino interaction simulation program library NEUT. Acta Phys. Pol. B **40**, 2477–2489 (2009), http://www.actaphys.uj.edu.pl/fulltext?series=Reg&vol=40&page=2477
21. S. Fukuda et al. (Super-Kamiokande Collaboration), The super-kamiokande detector. Nucl. Instr. Methods Phys. Res. Sect. A Accel. Spectr. Detec. Assoc. Equ. **501**(2–3), 418–462 (2003). doi:10.1016/S0168-9002(03)00425-X
22. Kamioka Observatory, ICRR (Institute for Cosmic Ray Research), The University of Tokyo, figure taken from, http://www-sk.icrr.u-tokyo.ac.jp/sk/detector/image-e.html
23. S. Kasuga et al., A study on the $e\mu$ identification capability of a water cherenkov detector and the atmospheric neutrino problem. Phys. Lett. B **374**(1), 238–242 (1996). doi:10.1016/0370-2693(96)00138-4
24. Y. Ashie et al. (Super-Kamiokande Collaboration), Measurement of atmospheric neutrino oscillation parameters by super-kamiokande I. Phys. Rev. D **71**, 112005 (2005). doi:10.1103/PhysRevD.71.112005
25. Kamioka Observatory, ICRR (Institute for Cosmic Ray Research), The University of Tokyo, figures taken from, http://www.ps.uci.edu/tomba/sk/tscan/compare_mu_e
26. M. Nakahata et al., Atmospheric neutrino background and pion nuclear effect for KAMIOKA nucleon decay experiment. J. Phys. Soc. Japan **55**(11), 3786–3805 (1986). doi:10.1143/JPSJ.55.3786
27. J.A. Formaggio, G.P. Zeller, From eV to EeV: neutrino cross sections across energy scales. Rev. Mod. Phys. **84**, 1307–1341 (2012). doi:10.1103/RevModPhys.84.1307
28. D. Griffiths, *Introduction to Elementary Particles*, 2nd revised edn. (Wiley, New Jersey, 2008)
29. D. Casper, The nuance neutrino physics simulation, and the future. Nucl. Phys. B Proc. Suppl. **112**(1), 161–170 (2002). doi:10.1016/S0920-5632(02)01756-5
30. C. Wilkinson et al., Testing charged current quasi-elastic and multinucleon interaction models in the NEUT neutrino interaction generator with published datasets from the MiniBooNE and MINERνA experiments. Phys. Rev. D **93**, 072010 (2016). doi:10.1103/PhysRevD.93.072010
31. A. Bercellie et al., Cross section parameters for 2014 oscillation analysis, Technical Report 192, T2K (2014), http://www.t2k.org/docs/technotes/192
32. A. Bercellie et al., Implementation of additional NIWG cross section parameterizations for 2014 analyses, Technical report 193, T2K (2014), http://www.t2k.org/docs/technotes/193
33. M. Dunkman et al., Updated recommendation of the 2014-5 oscillation parameters, Technical report 265, T2K (2016), http://www.t2k.org/docs/technotes/265
34. R.A. Smith, E.J. Moniz, Neutrino reactions on nuclear targets. Nucl. Phys. B **43**, 605 (1972). doi:10.1016/0550-3213(75)90612-4; (Erratum: Nucl. Phys. B **101**, 547 (1975))
35. O. Benhar, A. Fabrocini, Two-nucleon spectral function in infinite nuclear matter. Phys. Rev. C **62**, 034304 (2000). doi:10.1103/PhysRevC.62.034304
36. J. Nieves, I.R. Simo, M.J.V. Vacas, Inclusive charged-current neutrino-nucleus reactions. Phys. Rev. C **83**, 045501 (2011). doi:10.1103/PhysRevC.83.045501

37. R. Gran, J. Nieves, F. Sanchez, M.J.V. Vacas, Neutrino-nucleus quasi-elastic and 2p2h interactions up to 10 GeV. Phys. Rev. D **88**, 113007 (2013). doi:10.1103/PhysRevD.88.113007
38. D. Rein, L.M. Sehgal, Neutrino-excitation of baryon resonances and single pion production. Ann. Phys. **133**(1), 79–153 (1981). doi:10.1016/0003-4916(81)90242-6
39. K.M. Graczyk, J.T. Sobczyk, Form factors in the quark resonance model. Phys. Rev. D **77**, 053001 (2008). doi:10.1103/PhysRevD.77.053001
40. C. Wilkinson et al., Reanalysis of bubble chamber measurements of muon-neutrino induced single pion production. Phys. Rev. D **90**, 112017 (2014). doi:10.1103/PhysRevD.90.112017

Chapter 4
Bayesian Inference and the Markov Chain Monte Carlo Method

This thesis describes a Bayesian oscillation analysis, which uses a Markov Chain Monte Carlo (MCMC) method to extract estimates of the oscillation parameters from T2K neutrino and antineutrino data. In this chapter the theory behind the MCMC method will be presented, as well as more specific details about how the method is used in this analysis and how parameter estimates can be extracted from the fit results. Much of the information in this chapter is summarised from [1–5].

The analysis presented in this thesis relies on a simultaneous fit to near and far detector data. We consider—in addition to the six oscillation parameters that we wish to measure (three mixing angles, one CP-violating phase, and two mass-squared splittings)—743 systematic parameters: 26 to describe cross-section uncertainties, 100 to describe bin-normalisation uncertainties due to the neutrino flux, 580 to describe bin-normalisation uncertainties related to the ND280 detector, and 37 parameters to describe bin-normalisation uncertainties due to the SK detector response. A multi-dimensional probability distribution for the model parameters given the data is found, with one dimension for each parameter in the fit (in this case, 749). This is then integrated over all but the parameters of interest in order to make estimates of those parameters, a process known as marginalisation.

Monte Carlo methods rely on generating samples from the required probability distribution. These samples—in the limit of large numbers—can be used to approximate properties of the full distribution (for example, the true mean of a distribution can be approximated by the sample mean of the Monte Carlo sample). MCMC is a variant of the basic Monte Carlo method, in which the samples are generated by running a cleverly-constructed Markov chain.

The general approach to parameter estimation in the Bayesian framework is described in Sect. 4.1. Monte Carlo methods in general, and MCMC in particular, are then discussed in Sects. 4.2 and 4.3 respectively. Finally, the specific methods used in this thesis to extract parameter estimates and their uncertainties, as well as quantify how well a given model describes the data and compare it to other models, are presented in Sect. 4.4

K.E. Duffy, *First Measurement of Neutrino and Antineutrino Oscillation at T2K*, Springer Theses, DOI 10.1007/978-3-319-65040-1_4

4.1 Bayesian Inference

In the Bayesian world view there is no fundamental difference between observables and parameters of a statistical model, so in order to make inferences about model parameters we must find a full probability model of both the data and the model parameters. This is known as the joint probability distribution. If D represents the observed data and $\vec{\theta}$ all model parameters, we can write the joint probability distribution as $P(D, \vec{\theta})$. This is equal to a product of two parts: a prior distribution for the model parameters, $P(\vec{\theta})$, and a likelihood, $P(D|\vec{\theta})$.

$$P(D, \vec{\theta}) = P(D|\vec{\theta})P(\vec{\theta})$$

The prior distribution, $P(\vec{\theta})$, contains all previous knowledge about the parameters in the model. This could be a previous measurement (for example constraining a certain parameter to a best-fit value from external measurements with a corresponding Gaussian uncertainty), and may also include correlations between the parameters. $P(D|\vec{\theta})$ is the likelihood of measuring some data D assuming some set of parameter values given by the vector $\vec{\theta}$. In this analysis the likelihood is calculated by comparing a Monte Carlo prediction using a specific set of parameter values to the binned data, and evaluating the Poisson likelihood for each bin.

The aim of the analysis is to find the probability for each parameter to have a certain value, given the data: $P(\vec{\theta}|D)$. For example, we may wish to find the probability that $\delta_{CP} = 0$ given the information in the T2K data. This can be found from the joint probability distribution by applying Bayes' theorem:

$$P(\vec{\theta}|D) = \frac{P(D|\vec{\theta})P(\vec{\theta})}{\int P(D|\vec{\theta})P(\vec{\theta})d\vec{\theta}} \tag{4.1}$$

$P(\vec{\theta}|D)$ is called the posterior distribution (often just called the posterior)—the probability distribution for the model parameters $\vec{\theta}$ given the observed data D. Constructing the posterior distribution is the aim of all Bayesian analyses; from the posterior distribution it is easy to extract parameter estimates and other information about the model (as discussed in Sect. 4.4).

Note that we often cannot easily evaluate the integral in the denominator of Eq. 4.1, which is the integral of the joint probability distribution over all possible models and parameters. In this case we can only say that $P(\vec{\theta}|D)$ is proportional to $P(D|\vec{\theta})P(\vec{\theta})$. However, in Sect. 4.4 we shall see that it is sufficient in most cases (in fact, in all cases in this analysis) to know the posterior up to some normalisation constant.

4.2 Monte Carlo Methods

The basic premise behind all Monte Carlo methods is the same: produce samples from the desired distribution and then evaluate properties of the samples to use as approximations of the properties of the distribution. These properties could be parameter estimates, regions of high probability, or integrals over the distribution.

Consider this example of Monte Carlo integration from [3], shown in Fig. 4.1. We want to evaluate the area between the two grey curves, but can't analytically integrate either of the curves. Instead we turn to a Monte Carlo method: scatter points uniformly over the entire area, see what fraction of points lie between the curves, and multiply that fraction by the total area over which the points were scattered.

Now instead imagine we want to evaluate the integral of some function over the area inside the curves (labelled 'A'):

$$I = \int_A exp[-0.1(x^2 + y^2)]dxdy \tag{4.2}$$

This can be done in the same way: scatter random points across the entire area, and if the point lies in A, evaluate the function of the integrand. If not, evaluate it as zero. The integral I is then equal to the average value of the function multiplied by the total area over which the points were scattered.

This simple example demonstrates that Monte Carlo methods can be very powerful, particularly in problems which cannot be solved analytically. It is not necessary to know the full shape of a distribution in order to evaluate its properties using a

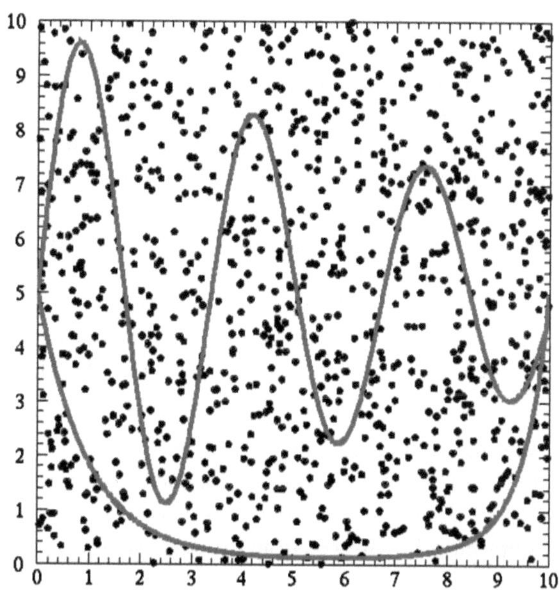

Fig. 4.1 A demonstration of integration by Monte Carlo. Figure from [3]

Monte Carlo method, but simply to be able to calculate its value at any given point in parameter space. However, methods such as the integration method presented here are very inefficient: if the area over which the points are to be scattered is not chosen well, we could waste a lot of time scattering points outside A which will not contribute to the integral I.

4.3 Markov Chain Monte Carlo

Markov Chain Monte Carlo (MCMC) is a particular Monte Carlo technique which uses a Markov chain to 'choose' points at which to evaluate the distribution—a much more efficient method than the random sampling demonstrated in Fig. 4.1. The Markov chain conducts a semi-random walk through parameter space, building up a collection of discrete points whose density is proportional to that of the desired distribution. The samples produced by the Markov chain are not independent, but do correctly cover the space of the distribution in the right proportions—a sufficient condition for a Monte Carlo method to work.

Note that, although MCMC is more efficient than other Monte Carlo methods, it is still very computer-intensive. The analysis described in this thesis makes use of GPUs to parallelise and therefore speed up the computation.

A Markov chain is a stochastic process defined such that the position at the end of each step $\vec{x}_i$ depends only on the previous position $\vec{x}_{i-1}$ and not on the further history of the chain ($\vec{x}_0$, $\vec{x}_1$, and so on). The space in which the Markov chain exists has N dimensions, where N is the number of parameters of the distribution that we wish to sample (in this thesis, this is the posterior probability distribution, which depends on 749 parameters). Each discrete point to which the chain steps can be defined by a vector $\vec{x}$ in this N-dimensional space.

Certain Markov chains have the property that as time increases the chain will gradually 'forget' its initial state, known as 'convergence'. Let $P^n(\vec{x}_n|\vec{x}_0)$ be the distribution of $\vec{x}_n$ (i.e. the distribution from which the n'th step is sampled) given some starting point $\vec{x}_0$. When the chain has 'forgotten' its initial state, $P^n(\vec{x}_n|\vec{x}_0)$ will converge to a unique stationary distribution, meaning that it is independent of the step number n. Once the chain has converged on this stationary distribution, all subsequent sampled points $\vec{x}_n$ will look like samples from that distribution.

The theory behind which chains show convergence is beyond the scope of this thesis, but an in-depth description is given in Chap. 4 of [1] and a summary in [2]. In short: it is necessary and sufficient that the chain satisfies the following three conditions, known as 'regularity conditions':

- **Irreducibility**: From all possible initial states in parameter space, the chain must have non-zero probability to reach all other potential states.
- **Recurrence**: Once the stationary distribution has been reached, then all subsequent steps must be samples from the same stationary distribution.

- **Aperiodicity**: The chain must not be periodic (i.e. cannot move periodically back and forth between two or more states).

If these conditions are satisfied then the chain will eventually converge on a stationary distribution.

This, then, gives us an analysis method: construct a Markov chain which has the posterior $P(\vec{\theta}|D)$ as its stationary distribution and satisfies the regularity conditions, and once it has converged on that distribution, use the output to estimate the model parameters $\vec{\theta}$. The only question that remains is how to construct a Markov chain with the correct stationary distribution, which is answered by using a technique developed by Metropolis et. al. [6].

4.3.1 *The Metropolis-Hastings Algorithm*

Not all Markov chains have stationary distributions, as mentioned above, but all Markov chains used for MCMC must. The Metropolis-Hastings algorithm provides a method to construct a chain that satisfies the regularity conditions and generates samples from a specified stationary distribution. It was first devloped by Metropolis et. al. in 1953 [6] and generalised by Hastings in 1970 [7].

A flow chart depicting the Metropolis-Hastings algorithm is given in Fig. 4.2. The method is as follows: at the end of step n, the chain is at some point in parameter space $\vec{x}_n$. To choose the next state $\vec{x}_{n+1}$, a candidate point $\vec{y}_n$ is generated randomly from some proposal distribution $q(\vec{y}_n|\vec{x}_n)$. Note that this proposal function may depend on the current point $\vec{x}_n$, but not on the further history of the chain. This step must be evaluated and either accepted as the next step in the chain (by setting $\vec{x}_{n+1} = \vec{y}_n$) or rejected (in which case we keep the current step: $\vec{x}_{n+1} = \vec{x}_n$). To decide whether to reject or accept the step, calculate the acceptance probability α:

$$\alpha(\vec{x}_n, \vec{y}_n) = \min\left(1, \frac{P(\vec{y}_n|D)q(\vec{x}_n|\vec{y}_n)}{P(\vec{x}_n|D)q(\vec{y}_n|\vec{x}_n)}\right)$$

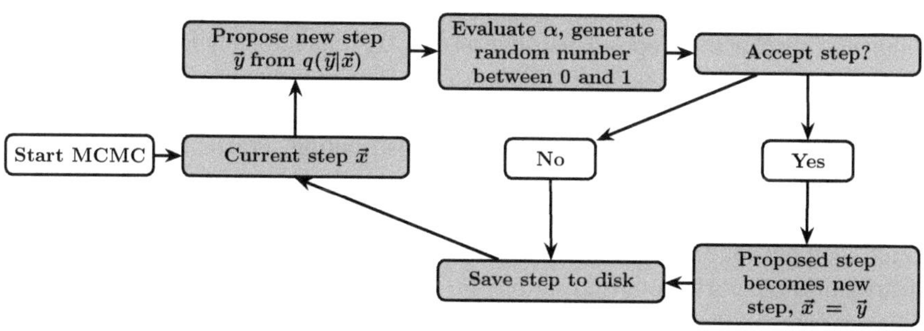

Fig. 4.2 The Metropolis-Hastings algorithm

where $P(\vec{x}_n|D)$ is the posterior probability, $P(\vec{\theta}|D)$, at some set of parameter values $\vec{\theta} = \vec{x}_n$ (but it could be any distribution from which we want the chain to sample). The analysis presented in this thesis uses a variant of the original Metropolis algorithm (before it was generalised by Hastings), which considers only symmetric proposal functions $q(\vec{x}_n|\vec{y}_n) = q(\vec{y}_n|\vec{x}_n)$. In this case the acceptance probability simplifies to the ratio of posterior probabilities for the two points:

$$\alpha(\vec{x}_n, \vec{y}_n) = \min\left(1, \frac{P(\vec{y}_n|D)}{P(\vec{x}_n|D)}\right)$$

A random number is then generated from a uniform distribution between 0 and 1. If this number is smaller than α then the step is accepted, $\vec{x}_{n+1} = \vec{y}_n$. If not, the step is rejected and the chain stays at the current step, $\vec{x}_{n+1} = \vec{x}_n$. This is often called Metropolis rejection—it is similar to the "rejection sampling" Monte Carlo technique, but in rejection sampling we sample repeatedly until some proposal is accepted (so the algorithm always produces a new value of the state). In the Metropolis-Hastings algorithm, we either accept the new state $\vec{y}_n$ (with probability α), or the new state stays the same as the old state $\vec{x}_n$.

If the proposed step $\vec{y}_n$ has a larger posterior probability than the current step $\vec{x}_n$ then $\alpha = 1$ and the step is always accepted. If it has a smaller probability than the current step then it may still be accepted, with acceptance probability α. Steps to points with higher posterior probability are more likely to be accepted than steps to points with low posterior probability.

Once the new step has been chosen (either $\vec{x}_{n+1} = \vec{y}_n$ or $\vec{x}_n$) it is saved to disk and the process of choosing a new step is repeated. In this way the algorithm builds up a distribution of discrete points with more points in regions of higher probability and fewer points in regions of lower probability. The density of points in the chain output will be proportional to the posterior probability density.

4.3.2 Proposal Function and Step Size

It can be shown (for example, in Chap. 1 of [1]) that the Metropolis-Hastings algorithm ensures that the proposal distribution $q(\vec{y}_n|\vec{x}_n)$ can have any form and the chain will always have a stationary distribution $P(\vec{x}|D)$: the stationary distribution of the chain is completely independent of the proposal function. However, this only shows that *eventually* the chain will settle on that distribution. Practically speaking, we want it to converge as quickly as possible because only the steps from after the chain has converged are samples from the posterior distribution. We also want the chain to 'mix' well once it has converged—this means that the steps should cover the whole distribution and not be focussed entirely on one area, or be too slow to move between areas of parameter space.

Figure 4.3 shows the accepted steps from 500 steps of three different MCMC chains. All have the same stationary distribution, a Gaussian distribution centered

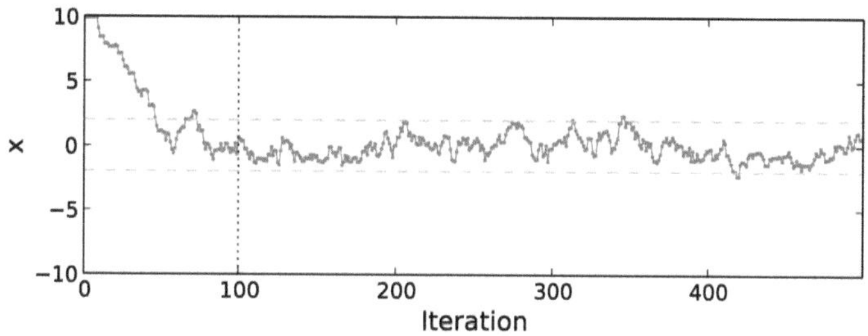

(a) $q(\vec{y}_n|\vec{x}_n) = N(\vec{x}_n, 0.5)$

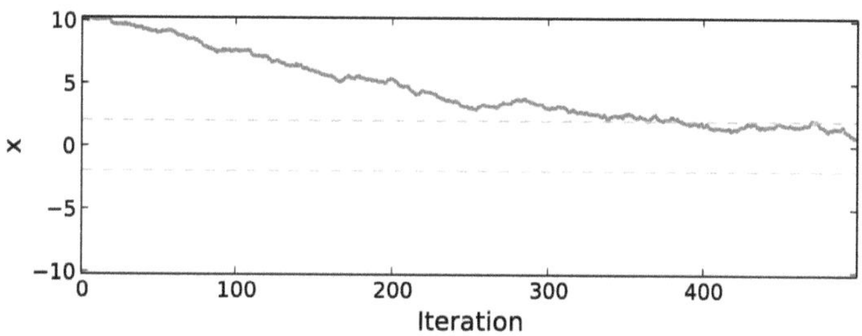

(b) $q(\vec{y}_n|\vec{x}_n) = N(\vec{x}_n, 0.1)$

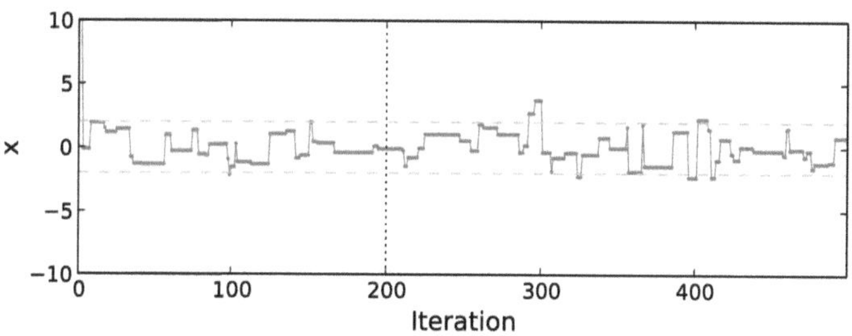

(c) $q(\vec{y}_n|\vec{x}_n) = N(\vec{x}_n, 10.0)$

Fig. 4.3 500 Iterations from MCMC chains using the Metropolis-Hastings algorithm with stationary distribution $N(0, 1)$, where $N(a, b)$ is a Gaussian distribution centered on a with width b. The *dashed grey lines* indicate $\pm 2\sigma$ of the stationary distribution, and *dotted lines* indicate the number of steps cut for burn-in (see Sect. 4.3.3). Figure created using MCMC software developed by A. Sztuc

on 0 with width 1, and all have proposal functions $q(\vec{y}_n|\vec{x}_n)$ which are Gaussian distributions centered on the current point $\vec{x}_n$, but with different widths. We can define the "step size" for each chain as the width of the Gaussian proposal function. When the step size is increased, the width of the proposal distribution is increased, which results in an increase in the 1D range in which the chain can step for a single parameter.

All three chains have the same stationary distribution to which they will eventually converge. Figure 4.3a has proposal function $N(\vec{x}_n, 0.5)$ (i.e. a Gaussian distribution centered on the current point $\vec{x}_n$ with width 0.5). The chain converges very quickly from a fairly extreme starting value, and then mixes quickly. Figure 4.3b has step size 0.1, which is too small—the chain hasn't really 'forgotten' its starting value or begun exploring the stationary distribution after 500 iterations. When the step size is small, every step has a high acceptance probability because $P(\vec{y}_n|D) \simeq P(\vec{x}_n|D)$. However, the chain does not mix well—it would need to be run for a very long time to get a good sample of the distribution across the whole parameter space. Figure 4.3c, on the other hand, has a very large step size. It often proposes a step out in the tails of the distribution where $P(\vec{y}_n|D)/P(\vec{x}_n|D)$ is small, leading to a low probability of acceptance. Because of this the chain frequently doesn't move. As in chain (b), a large number of steps would be needed to get a reliable estimate of the stationary distribution over the whole space.

It is clear that the proposal function (and step size) has a large effect on the number of steps needed to obtain a sample of points that reliably approximates the actual posterior density. If the step size is too small, many points are accepted but they are concentrated in a small space and it will take the chain a long time to move from one part of the distribution to another. In this case the chain is called "highly correlated"—it cannot step far enough so only explores its local density. If the step size is too large, many of the proposed points will be rejected so the chain will often not move for a long time.

Step size tuning is a notoriously difficult feature of the MCMC technique, and it is very important to get right. In this analysis, the step size for each group of parameters (e.g. flux, cross section, detector response) is tuned by hand. Some guidance comes from [8], a paper which discusses techniques to determine the optimum step size for MCMC when the proposal function is Gaussian and centered on the current step. The authors use a dimensionless measure of efficiency of an MCMC chain, E, defined by comparing the sample mean variance from the chain with that of an 'ideal' chain:

$$E = \lim_{N\to\infty} \frac{\sigma_0^2/N}{\sigma_{\bar{x}}^2(N)}$$

where $\sigma_{\bar{x}}^2(N)$ is the sample mean variance of an MCMC chain with N steps. σ_0^2 is the variance of the underlying distribution, meaning that σ_0^2/N is the sample mean variance of an ideal chain (i.e. an MCMC chain sampling from the underlying distribution with no step-to-step correlations) of length N. Maximising this efficiency is equivalent to minimising the number of steps needed to obtain a sufficient sample from the posterior distribution. Using this metric, the authors find that the ideal acceptance rate (proportion of proposed steps that are accepted) is ~45% for a 1D chain and approaches ~25% for higher-dimensional problems. However, the number of dimensions considered by the authors in [8] is considerably smaller than in this analysis, so this guidance can only be considered a rough estimate. The correct step size for each group of parameters must be estimated by examining the 'trace' of each parameter (the value of that parameter as a function of step number, as shown in Fig. 4.3), and the trace of the log-likelihood, to ensure that the chain converges quickly and mixes well.

The analysis described in this thesis uses a special case of the Metropolis-Hastings algorithm called the random-walk Metropolis algorithm, in which—as in the examples above—the proposal function is a multivariate Gaussian centred on the current state $\vec{x}_n$. Correlations between parameters (according to the priors discussed in Sects. 5.1 and 5.3) are included in the proposal function because when steps are proposed without correlations (i.e. the proposal function for each parameter is independent of the other parameters), a lot of steps are proposed that will have a small prior probability $P(\vec{\theta})$. These are unlikely to be accepted, so the efficiency of the MCMC will suffer. For example, the prior constraint used in this thesis on the systematic parameters describing the neutrino flux says that the flux should be highly correlated between neutrino and antineutrino mode beams (this is because in both cases the same proton beam impinges on the same target, and the only difference is the current in the magnetic focussing horns). Any proposed step with a high neutrino flux and a very low antineutrino flux will have a very low prior probability, leading to a low posterior probability $P(\vec{\theta}|D)$, and so is unlikely to be accepted. When prior correlations are included in the step proposal function, steps are more likely to be proposed near the region of high probability, and so more steps are accepted.

The step size for each group of parameters (oscillation, flux, cross-section, ND280 detector, and SK detector) is tuned by hand to ensure an acceptance probability of $\sim$25% as recommended in [8].

4.3.3 Burn-In

The MCMC chain typically may not start at a region of high probability (as in Fig. 4.3), and therefore must be allowed to explore the parameter space in order to reach the stationary distribution. Even if it does start at a region of high probability, the condition given in Sect. 4.3 for the chain to produce samples from the posterior distribution is that it must have 'forgotten' its starting point. Because of this condition, the first steps at the beginning of the chain (before it has converged on the stationary distribution) are thrown away. These steps are colloquially called the 'burn-in'.

The dotted line in sub-figures (a) and (c) of Fig. 4.3 shows the burn-in. In sub-figure (a) a 100-step burn-in is needed, and in sub-figure (c) the burn-in is increased to 200 steps because of the slow mixing of the chain. In sub-figure (b) the chain has not 'forgotten' its starting position even in 500 steps, so a much longer burn-in period would be needed.

Figure 4.4 shows the log-likelihood as a function of step number for one of the fits presented in this thesis (the T2K-only Asimov fit described in Sect. 5.5.1.1). The first 100,000 steps of a number of independent MCMC chains are shown overlaid on top of each other. The log-likelihood changes quickly in the first 20,000 steps, showing that the chain has not yet converged on the stationary distribution. After this period, the log-likelihood remains stable for the duration of the chain. The burn-in

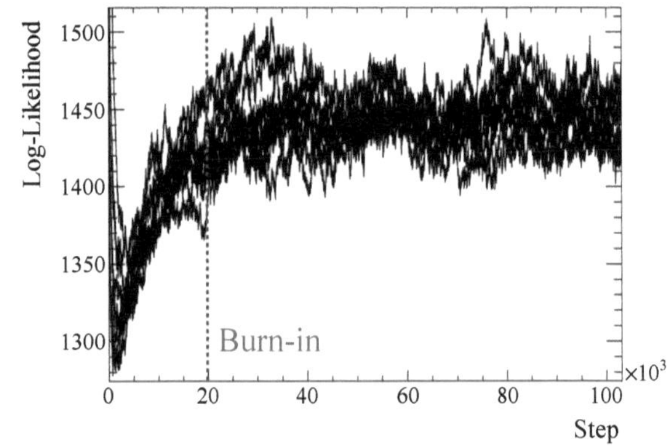

Fig. 4.4 Log-likelihood of each accepted step as a function of step number in the MCMC, for ten fit chains run independently for the T2K-only Asimov fit described in Sect. 5.5.1.1. Based on this figure, it was decided that the first 20,000 steps would be thrown away as burn-in, as shown by the dashed line

is dependent on the chain and fit conditions, so should be re-calculated for each fit. However, a burn-in period equal to 20,000 steps—as marked on Fig. 4.4—was found acceptable for all fits in this thesis.

4.4 Extracting Information from the Posterior

As we have established, MCMC is a very powerful tool for sampling from (and therefore producing an approximation of) a posterior distribution. This posterior distribution is the 'result' of the analysis, and in a perfect world Bayesians would prefer to publish the full distribution because it contains all the information about all model parameters. However, a 749-dimensional probability distribution is difficult to visualise and interpret, and not much use to anyone by itself. Therefore, once we have a sample from the posterior distribution, we must remove information so that the results can be interpreted with focus on the most important physical aspects.

4.4.1 Marginalisation of Nuisance Parameters

Although all model parameters stand on equal footing from the Bayesian perspective—there is no ideological difference between physics parameters and other model parameters—we are usually more concerned with a single parameter or subset of parameters. For example, in an oscillation analysis at T2K we want to report best-fit values and uncertainties for the interesting oscillation parameters $\sin^2\theta_{23}$, $\sin^2\theta_{13}$, Δm^2_{32}, and δ_{CP} without regards to the other model parameters. It is necessary, however, to ensure that the uncertainty that is introduced by these other parameters is taken into account properly when evaluating the parameters of interest.

Table 4.1 Joint probability distribution $p_{x,y}(i, j)$ for rolling two fair dice

j	i						
	1	2	3	4	5	6	$p_x(i)$
1	1/36	1/36	1/36	1/36	1/36	1/36	1/6
2	1/36	1/36	1/36	1/36	1/36	1/36	1/6
3	1/36	1/36	1/36	1/36	1/36	1/36	1/6
4	1/36	1/36	1/36	1/36	1/36	1/36	1/6
5	1/36	1/36	1/36	1/36	1/36	1/36	1/6
6	1/36	1/36	1/36	1/36	1/36	1/36	1/6
$p_y(j)$	1/6	1/6	1/6	1/6	1/6	1/6	

A 'nuisance parameter' is a model parameter that affects the probability distribution but which we don't care about for its own sake. For example, the calibration constant of a piece of apparatus: not the sort of thing we would report in the abstract of a paper, but important nonetheless, and will have an effect on the result. In this analysis that means parameters governing the neutrino flux, cross-section, and detector uncertainties, as well as the oscillation parameters to which T2K is not sensitive ($\sin^2\theta_{12}$ and Δm^2_{21}).

The Bayesian approach gives a very simple procedure for eliminating unwanted parameters from the posterior distribution, known as marginalisation. Essentially, this means integrating the posterior over all unwanted parameters so that it only spans a few dimensions (the interesting model parameters). This gives a posterior distribution that is easier to understand and doesn't contain too much extraneous information.

For example, consider the joint probability distribution for rolling two dice x and y, as given in [2]. Assuming both dice are fair, the joint probability distribution $p_{x,y}(i, j)$ is shown in Table 4.1. If we wish to find the probability that dice x will roll a 1, $p_x(1)$, then it is easy to see that the roll of dice y is irrelevant. We can find $p_x(i)$ for any value i from 1 to 6 by summing over all possible values of y, as in Table 4.1.

The same logic can be applied to an oscillation analysis. An MCMC chain is run, including all model parameters (with appropriate priors), to construct the joint posterior. This can then be integrated over the unwanted nuisance parameters to construct the "marginal posterior", which is now only a function of the parameters of interest:

$$P(\vec{\theta}_{poi}|D) = \frac{\int P(D|\vec{\theta}_{poi}, \vec{\theta}_n)P(\vec{\theta}_{poi}, \vec{\theta}_n)d\vec{\theta}_n}{\int P(D|\vec{\theta})P(\vec{\theta})d\vec{\theta}} \tag{4.3}$$

where $\vec{\theta}_{poi}$ are the parameters of interest and $\vec{\theta}_n$ are the nuisance parameters. This is equivalent to projecting the distribution on to a smaller number of dimensions. In this analysis the posterior is marginalised onto either one dimension (for a 1D point estimate or uncertainty), two dimensions (for a 2D point estimate and contour, for

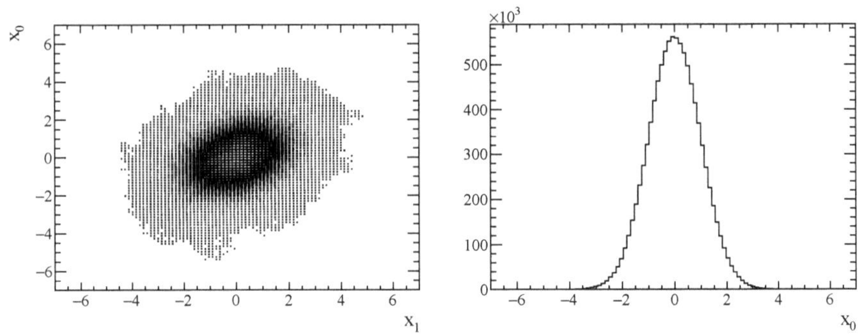

Fig. 4.5 *Left* A 2D correlated Gaussian distribution as a function of two parameters, x_0 and x_1. *Right* the same distribution, marginalised over x_1 to be a function of x_0 only

example in $\sin^2\theta_{23}$–Δm^2_{32} as in Fig. 4.7), or four dimensions (to find the best-fit point with respect to the four interesting oscillation parameters).

In practise, evaluating Eq. 4.3 is non-trivial, because for each nuisance parameter we have to add a dimension to the integral. This is one of the reasons MCMC is used for this analysis—it is very well-suited to evaluating high-dimensional integrals that could not be solved analytically.

While marginalisation is the correct way to deal with nuisance parameters in a Bayesian framework, it can produce some unintuitive results. If the nuisance parameters which are being marginalised have non-Gaussian distributions, or correlations with the parameters of interest, this can move the region of high probability in a marginal posterior probability distribution away from where one might expect it to be.

Consider the simple example shown in Fig. 4.5: a correlated Gaussian in two parameters, x_0 and x_1, centred on 0. If we consider this to be the posterior probability distribution of some two-parameter fit, the plot on the left in Fig. 4.5 shows the posterior as a function of both parameters, and the plot on the right shows the posterior as a function of only parameter x_0, with parameter x_1 marginalised. In this case the marginalised parameter x_1 has a Gaussian distribution, and the 1D marginal posterior peaks at $x_0 = 0$ (which is also the peak of the 2D distribution).

Now imagine that we impose a prior constraint on parameter x_1 that it must have positive value, as in Fig. 4.6. In this case the distribution of x_1 is no longer Gaussian. Now when we marginalise x_1 to find the 1D posterior probability distribution as a function of x_0, we find that it has its peak above 0. This is called a 'marginalisation effect': the most probable point of x_0 depends on the number of dimensions of the probability distribution. In the 2D distribution the peak is still at $x_0 = 0$, but in the 1D distribution it is shifted.

In this way, marginalisation can produce an apparent bias. A fit may be completely unbiased, in the sense that the full posterior probability distribution of a fake data fit has a maximum at the correct value of all parameters, but still appear to have a "bias" in the marginal probability distribution if the distribution in some of the

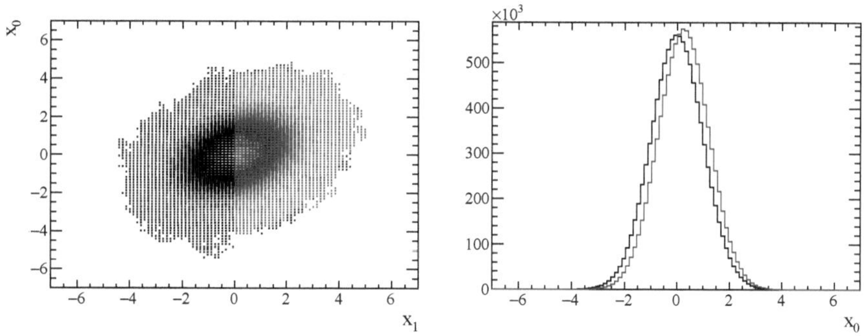

Fig. 4.6 *Left* A 2D correlated Gaussian distribution as a function of two parameters, x_0 and x_1. *Right* the same distribution, marginalised over x_1 to be a function of x_0 only. The *red* distribution has a cutoff at $x_1 = 0$, which produces a marginalisation bias in the 1D distribution. It is shown overlaid with the distribution from Fig. 4.5 for comparison

nuisance parameters is not Gaussian. This is very likely to happen in the oscillation fits presented in this thesis—the posterior distributions of many of the nuisance parameters are not Gaussian. We have no reason to expect oscillation parameters such as δ_{CP} to have a Gaussian distribution (in fact the distribution is very much non-Gaussian), and certain cross-section systematic parameters have physical limits which cause their posterior distributions to resemble a truncated Gaussian (as in the example above).

Marginalisation effects are seen in all of the Asimov fits used to validate the analysis presented in this thesis (shown in Sect. 5.5.1). It is important to examine these in order to understand what we expect from the data fit. However, this does not mean the fit is biased; it is a true effect in the marginal posterior distribution.

4.4.2 Point Estimation

The natural way to define a best-fit value from a posterior distribution is to use the mode of the distribution, equal to the most probable point. In the MCMC method, this requires us to find the region of highest density of a series of discrete points in parameter space.

The simplest approach is to use a histogram to collect the posterior density information into binned intervals. The bin containing the most points is then the region of highest density. This approach is easy to implement and widely used—in this thesis it is used when reporting the best-fit values of systematic parameters in 1 dimension (the posterior is projected into a 1D histogram and the 1D mode reported as the best fit), and when reporting 2D best-fit values for the oscillation parameters.

The drawback to this approach is that when the number of 'interesting' parameters becomes large, the number of bins in the N-dimensional histogram also increases, and bin statistics can become a problem. We need higher statistics (i.e. a larger number of steps in the MCMC output) to ensure that the bins are sufficiently populated, and that may not be practically possible. We could use wider bins, but that decreases the precision of the density estimation.

T2K is sensitive to four oscillation parameters, which means that we must find the mode of a 4D distribution. It is impractical to populate all bins sufficiently in 4D, so a kernel density estimation technique is used to turn the series of discrete points into a smooth continuous density function, the maximum of which can then be found using Minuit [9]. The kernel density estimation used in this thesis relies on a Gaussian 'kernel'. A single kernel is placed over each data point, and the sum of all kernels produces a smooth estimation of the density of points in the 4D marginal posterior distribution.

The width of these Gaussian kernels is known as the 'bandwidth', and it is very important that the bandwidth is not too small (which will lead to undersmoothing, and a resultant distribution which is spiky and difficult to interpret) or too large (which would cause oversmoothing, and mask the structure of the distribution). To ensure that the distribution is neither undersmoothed nor oversmoothed, an 'adaptive kernel density estimator' [10] is used, which adjusts the bandwidth to the local density of points: a large bandwidth is used in areas of low density, and a smaller bandwidth in areas of high density. More information about this method is given in Chap. 4 of [2].

The best-fit point may well depend on the number of dimensions onto which the posterior is marginalised, due to the marginalisation effects discussed in Sect. 4.4.1. For example, the best-fit point in $\sin^2\theta_{23}$–Δm^2_{32} 2D space (in which all other parameters, including $\sin^2\theta_{13}$ and δ_{CP} are marginalised) may differ from the best-fit values of $\sin^2\theta_{23}$ and Δm^2_{32} in 4D $\sin^2\theta_{23}$–Δm^2_{32}–$\sin^2\theta_{13}$–δ_{CP} space.

Figure 4.7 shows the 2D marginalised posterior distribution from the T2K-only Asimov fit presented in Sect. 5.5.1.1, as a function of $\sin^2\theta_{23}$ and Δm^2_{32}. The marginal posterior is presented as a shaded histogram, with darker shades representing larger posterior density (i.e. more probable regions). The 2D and 4D best-fit points are both shown, and it is clear that there is a marginalisation effect when marginalising over either or both of $\sin^2\theta_{13}$ and δ_{CP}, because the two best-fit points are not identical. In fact, the marginalisation effect is due to correlations between $\sin^2\theta_{23}$ and $\sin^2\theta_{13}$, discussed further in Sect. 5.5.1.1.

4.4.3 *Uncertainty on Point Estimates*

Once point estimates have been found, it is essential to also give an uncertainty on that estimate. This is done by constructing Bayesian credible intervals from the marginalised posterior distribution. A credible interval is defined as the interval in the marginal posterior that contains a certain fraction of the total posterior density. For example, the integral

$$\int P(\vec{\theta}|D)d\vec{\theta} = \alpha$$

will define a region that contains $\alpha \times 100\%$ of the total probability in the posterior distribution. If $\alpha = 0.9$ this corresponds to a 90% credible interval.

There are many ways to define credible intervals which are valid—the only constraint is that a 90% credible interval must contain 90% of the posterior probability. This analysis uses highest posterior density (HPD) credible intervals, which are defined such that any point inside the credible interval has a larger posterior density than any point outside the interval.

The procedure for constructing HPD credible intervals is as follows: first, a histogram of the marginal posterior must be produced in the desired number of dimensions (in this thesis, 1D credible intervals are used for all systematics and the interesting oscillation parameters, and 2D credible intervals for certain combinations of the interesting oscillation parameters). Bins are then added from the histogram in order of highest to lowest population until the bins counted so far contain $\alpha \times 100\%$ of the total posterior density. For example, to report a 90% credible interval bins are added until the counted bins contain 90% of the integral of the posterior.

The results of this thesis are typically presented as 68 and 90% credible intervals, in keeping with the precedent set by frequentist analyses (a 68% frequentist confidence level is equivalent to $\Delta\chi^2 = 1\sigma$). These are shown over the marginal posterior in 2D $\sin^2\theta_{23}$–Δm^2_{32} in Fig. 4.7.

Note that if the posterior distribution has multiple maxima the HPD credible interval construction can lead to a single credible interval being composed of multiple non-contiguous areas. This is the case in Fig. 4.7: only one 90% and one 68% credible

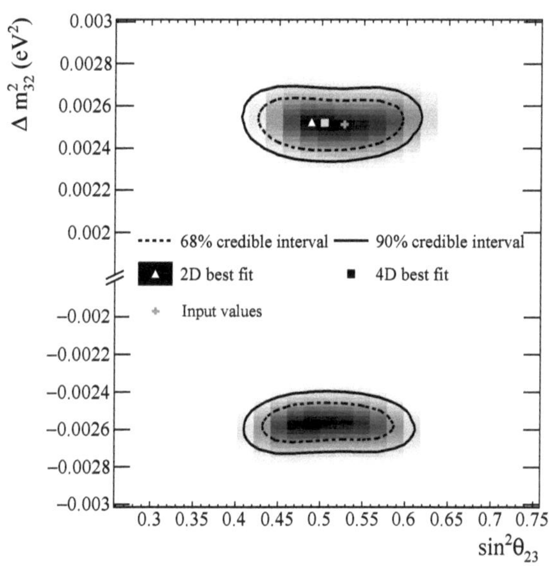

Fig. 4.7 2D marginal fit posterior in $\sin^2\theta_{23}$–Δm^2_{32} from the T2K-only Asimov fit (copied from Fig. 5.11 in Sect. 5.5.1.1). The 2D credible intervals, 2D best-fit point, and 4D best-fit point are also shown, as well as the input values, $\sin^2\theta_{23}$=0.528, and Δm^2_{32}= 2.509×10^{-3}

interval are shown, but both credible intervals contain two distinct areas in parameter space. The integral of the posterior density over both areas contained in the two '90% credible interval' curves is equal to 90% of the total posterior probability density.

Although the Bayesian credible interval is the correct "result" for this analysis, it is also possible to construct frequentist-style constant-$\Delta\chi^2$ intervals for the sole purpose of comparing validations and results to other analyses on T2K. T2K has two other oscillation analysis groups, both of which use a hybrid frequentist-Bayesian approach that marginalises over nuisance parameters and then produces frequentist confidence intervals (the similarities and differences between the three oscillation analysis groups on T2K are described in Appendix B).

Confidence intervals as a function of the oscillation parameters $\sin^2\theta_{23}$, $\sin^2\theta_{13}$, Δm^2_{32}, and δ_{CP} are constructed for comparisons; all of these parameters have flat priors, with the exception of $\sin^2\theta_{13}$. The prior used on these parameters is important because it is only possible to produce confidence intervals from a posterior distribution if the posterior can be interpreted as a likelihood. Recall Eq. 4.1: the posterior is proportional to the likelihood multiplied by the prior probability. Therefore, if the prior is flat in the parameters of interest, the marginal posterior can be interpreted as a marginal likelihood (in which all nuisance parameters are marginalised and may have non-flat priors).

There are two possible choices of prior on $\sin^2\theta_{13}$, and the analysis is repeated twice using both choices. The first is a flat prior in $\sin^2\theta_{13}$, which is sometimes referred to as a 'T2K only' or 'without reactor constraint' fit: only T2K data are used to constrain $\sin^2\theta_{13}$. The other choice is to use a Gaussian prior on $\sin^2 2\theta_{13}$, as measured by the reactor experiments and reported in the 2015 PDG [11]:

$$\sin^2 2\theta_{13} = 0.085 \pm 0.005 \tag{4.4}$$

This is often referred to as fitting 'with reactor constraint'. When the reactor constraint is used, the frequentist analyses add it to the likelihood as a penalty term. This produces a 'likelihood' that is equivalent to the Bayesian posterior with a prior. Therefore, confidence intervals produced by treating the posterior as a likelihood will be directly comparable to the confidence intervals produced by the frequentist T2K analyses.

To produce constant-$\Delta\chi^2$ confidence intervals, the binned marginal posterior is used and content of each bin is interpreted as a likelihood, $\mathcal{L}_{bin}$ (this is not strictly valid, but it is true that the content of each bin can be interpreted as proportional to a likelihood, and the constant of proportionality cancels in Eq. 4.5). The bin in which the posterior is largest is found, and the posterior density in that bin is defined as the maximum likelihood, $\mathcal{L}_{max}$. For each bin in the histogram, $\Delta\chi^2$ is then calculated:

$$\Delta\chi^2 = -2\ln\left(\frac{\mathcal{L}_{bin}}{\mathcal{L}_{max}}\right) \tag{4.5}$$

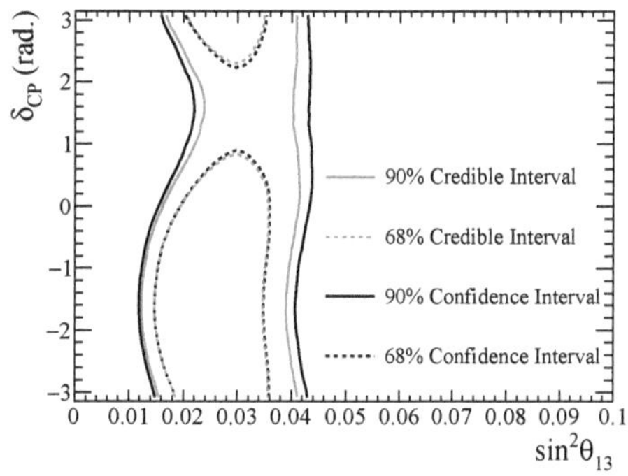

Fig. 4.8 2D $\sin^2\theta_{13}$–δ_{CP} credible intervals from the T2K-only Asimov fit described in Sect. 5.5.1.1 with constant-$\Delta\chi^2$ confidence intervals overlaid

Once the $\Delta\chi^2$ distribution is found, confidence intervals are produced by drawing contours at the critical $\Delta\chi^2$ values: $\Delta\chi^2 = 1$ for 68% and $\Delta\chi^2 = 2.71$ for 90% confidence intervals in 1D, and $\Delta\chi^2 = 2.30$ for 68% and $\Delta\chi^2 = 4.61$ for 90% confidence intervals in 2D.

Credible intervals and confidence intervals may often look similar, but their construction and interpretation are completely different, so there is no reason to expect them to be the same unless the posterior distribution is exactly Gaussian. The interpretation of a 90% confidence interval is that, if the experiment were to be repeated a large number of times, 90% of those times the measured best-fit point would lie within the 90% confidence interval. A 90% credible interval, however, simply means that—given *this* experiment—there is 90% probability that the true parameter value lies in the 90% credible interval. One of the most significant effects of this difference in interpretation is that the coverage study, a common frequentist test in which many fake data sets are fit to verify that the best-fit point of 90% of the fake data sets does indeed fall within the 90% confidence interval, is not applicable to credible intervals. A Bayesian 90% credible interval contains 90% of the posterior probability for a given experiment by construction.

Figure 4.8 shows the credible intervals and confidence intervals in 2D $\sin^2\theta_{13}$–δ_{CP} space from the T2K-only Asimov fit described in Sect. 5.5.1.1. The posterior distribution is non-Gaussian in both $\sin^2\theta_{13}$ and δ_{CP}, and we see a noticeable difference between the 90% credible intervals and confidence intervals, even though both are constructed from the same posterior distribution.

4.4.4 Model Comparison

So far we have considered the Bayesian approach (and the approach used in this thesis) to parameter estimation. This is essentially asking the question: 'within the context of a specific model, what parameter values are most consistent with the data?'. It is also useful to ask whether the model itself is even correct, a process known as hypothesis testing. The Bayesian framework gives a very easy method for answering this question, because in the Bayesian language all problems are hypothesis tests. At its most basic, parameter estimation simply amounts to assigning a degree of credibility to the hypothesis "x is between 5 and 5.01".

Recall that a parameter's posterior density is proportional to the prior multiplied by the marginal likelihood (when all other parameters are marginalised). In a similar way, the posterior probability for an entire model is proportional to the prior probability for that model multiplied by the global marginal likelihood (where all parameters are marginalised).

In the Bayesian framework, two models can be compared by calculating the posterior odds ratio for the models. The ratio of posterior odds for two models M_i and M_j is given by:

$$O_{ij} = \frac{P(M_i|D)}{P(M_j|D)}$$

which is the ratio of model M_i being favoured over M_j. Applying Bayes' theorem as in Eq. 4.1, we find

$$O_{ij} = \frac{P(D|M_i)P(M_i)}{P(D|M_j)P(M_j)} \equiv \frac{P(M_i)}{P(M_j)} B_{ij} \tag{4.6}$$

where B_{ij} is called the 'Bayes factor'—the factor by which we must multiply the prior odds of two hypotheses to find the posterior odds. The posterior odds can be interpreted as how much more likely model M_i is to be true than model M_j when considering all knowledge (prior probability densities and the evidence from the data). The Bayes factor, then, should be interpreted as the weight of evidence coming from the data (i.e. how we should update our opinion about models M_i and M_j from our prior knowledge once we have looked at the data). Equation 4.6 shows that the Bayes factor is simply equal to the ratio of likelihoods of the two models.

The Bayes factor was originally proposed by Jeffreys [12], who presented a scale through which the Bayes factor should be interpreted. This is given in Table 4.2, and shows a 'strength' of evidence for model M_i, as assigned by Jeffreys based on the value of the Bayes factor. This is somewhat arbitrary, and in practise the Bayes factor itself should always be presented and explained such that the reader can come to their own conclusions about the evidence presented by the data. To assist with this, a fourth column has been added to the table, labelled 'Degree of belief in model M_i'. The 'degree of belief' is calculated by assuming that there are only two possible models M_i and M_j, which between them span all possibilities:

$$P(M_i|D) + P(M_j|D) = 1$$

We also assume that the two models have equal priors $P(M_i) = P(M_j)$, in which case the Bayes factor is equal to the ratio of posterior odds: $B_{ij} = \frac{P(D|M_i)}{P(D|M_j)} = \frac{P(M_i|D)}{P(M_j|D)}$. Combining this constraint with the one above gives two equations in two unknowns

Table 4.2 The Jeffreys scale for interpretation of the Bayes factor $B_{ij} = \frac{P(D|M_i)}{P(D|M_j)}$ [12]. A fourth column has been added, showing the degree of belief in model M_i assuming that the only possible models are M_i and M_j, that both models have equal priors, and that $P(M_i|D) + P(M_j|D) = 1$

B_{ij}	$\log_{10} B_{ij}$	Strength of evidence for model M_i	Degree of belief in model $M_i(\%)$
<1	<0	Data prefers model M_j	<50
1–3.16	0–0.5	Preference for model M_i is weak	50–76
3.16–10	0.5–1	Preference for model M_i is substantial	76–91
10–100	1–2	Preference for model M_i is strong	91–99
>100	>2	Preference for model M_i is decisive	>99

and allows us to calculate $P(M_i|D)$. This is, of course, not generally applicable in the way that the Jeffreys scale is, as it requires that there are only two possible models which have equal prior probabilities. However, it is hoped that this will help the reader gain a more intuitive sense of the interpretation of a Bayes factor.

The frequentist approach to hypothesis testing relies on calculating a p-value, which represents the probability that—if a given model (the 'null hypothesis') is true—we would obtain data as anomalous or more anomalous than the data we have observed.

There is no universal correspondence between a Bayes factor and a p-value, due to a fundamental difference in how the two should be interpreted. The p-value informs the analyser how likely it is that a particular dataset has arisen by chance if a given null hypothesis is true; it is not the probability that the null hypothesis is true. Therefore it can only be used to reject hypotheses, and provides no evidence in favour of the alternative. The Bayes factor, on the other hand, can be interpreted as the evidence from the data that one of the hypotheses is true. It places the two models on equal footing and allows the data to generate evidence for either one (an important consequence of this is that, unlike the p-value, the Bayes factor can show evidence *for* the null hypothesis over the alternative).

Whether a p-value or a Bayes factor is most appropriate depends on the question we wish to answer. When we perform goodness-of-fit tests, where the question is "how well does this model fit my data?" or "how likely is it that I would see this data if this model were correct?" then the p-value is the ideal statistic. Indeed, that is what is used for goodness-of-fit tests in this analysis (as described in Sect. 4.4.6).

However, questions such as "which of these two models fits the data best?" or "what are the betting odds that this hypothesis is true compared to that one?" are fundamentally Bayesian questions and so the best way to answer them is by calculating a Bayes factor. The question of the neutrino mass hierarchy, in particular, is well-suited to a Bayesian answer because there is no obvious 'null' hypothesis to reject—we simply want to compare two models (normal and inverted hierarchy) and see which one is preferred by the data.

Bayes factors are calculated for two purposes in this thesis: as a measure of how much the data favours the normal over inverted mass hierarchy (encoded by the sign of Δm^2_{32}) in the joint ν+$\bar{\nu}$ analysis presented in Chaps. 5 and 6, and as a measure of evidence for $\bar{\nu}_e$ appearance in Appendix A.2. In both cases equal priors are used for both hypotheses, so the posterior odds ratio is simply equal to the Bayes factor.

4.4.5 Posterior Predictive Method

It is often desirable in the T2K oscillation analysis to present a best-fit energy spectrum for the SK data samples as a function of reconstructed neutrino energy, which can be easily compared by eye to the data. We cannot simply take the 'best-fit' point with respect to all fit parameters, because that would require us to find the mode of the 749-dimensional posterior distribution—to do so accurately would require an extremely large number of steps and be computationally prohibitive. Instead, a 'posterior predictive' method is used, which essentially marginalises over all parameters, oscillation included, to find the posterior probability distribution as a function of reconstructed energy.

The posterior predictive method is as follows: we wish to create a 'best-fit' energy spectrum, which will rely on some binning in reconstructed neutrino energy. To do this, some number N of discrete points are selected randomly from the fit posterior (in this thesis, all posterior predictive spectra are produced with $N = 2500$). Each point will be at a particular location in parameter space, and have some values for each of the model parameters (oscillation, neutrino flux, cross-section, ND280 detector, and SK detector). Because there are more points in the regions of higher posterior density, those parameter values close to regions with larger posterior probability are more likely to be selected.

We then consider a single bin in the energy spectrum. For each of the N selected points, we calculate the predicted event rate in that bin, assuming the parameter values defined by that point. The combination of all N points creates a distribution of event rates in the given bin. A Gaussian is fitted around the peak of this distribution, and the mean of the Gaussian taken to be the predicted value for that bin.

This is repeated for every bin in reconstructed neutrino energy, using the same N points every time, in order to construct the 'best-fit' energy spectrum. Note that—because each bin is treated independently—the best-fit spectrum produced by this method may not be a spectrum that it is possible to produce for some set of 'best-fit' parameters.

Figure 4.9 shows the best-fit reconstructed energy spectrum produced in this way for the SK FHC 1R$_\mu$ sample in the T2K-only data fit (the data fit is described in Chap. 6, and this figure is copied from Fig. 6.8a). The shaded 2D histogram shows the result of sampling 2500 points from the fit posterior, the dashed line shows the best-fit spectrum, and the points show the data in this SK sample. Note that the binning in this case is not the same as the binning used for the data fit—a coarser binning was chosen in order to reduce the effect of statistical variations between bins.

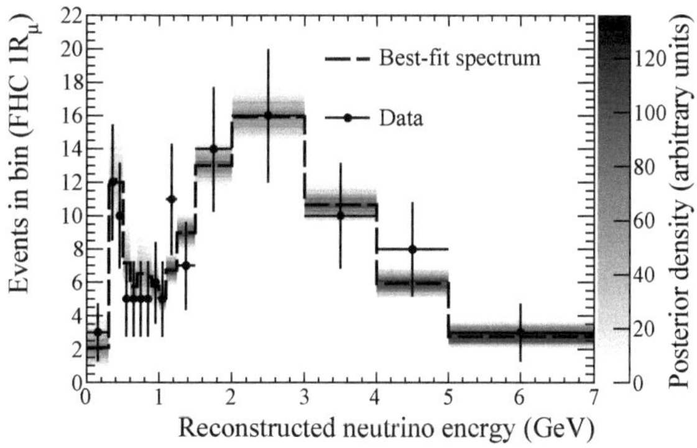

Fig. 4.9 Best-fit reconstructed neutrino energy spectrum for the SK FHC 1R$_\mu$ sample after the T2K-only data fit (copied from Fig. 6.8a in Chap. 6). Data points (with statistical uncertainties) are shown in *black*, the best-fit reconstructed energy spectrum is shown with a *black dashed line*, and the result of sampling the fit posterior 2500 times is shown as the shaded 2D histogram. A coarser binning is used than for the data fit, and the reconstructed energy range is cut off at 7 GeV because no data are found at higher energies (although the fit region extends to 30 GeV)

4.4.6 Goodness-Of-Fit

Once a fit to data has been completed, it is natural to ask how good the fit is. This is done using a goodness-of-fit test, which is a special kind of hypothesis test. Unlike the tests described in Sect. 4.4.4, in this case the question is not "does model A fit the data better than model B?" but simply "does this model fit the data?". Therefore there is no need to specify an alternate hypothesis—in essence, the 'alternate hypotheses' considered are the set of all possible statistical variations of the data.

In this thesis the goodness-of-fit test relies on a p-value based on the χ^2 statistic, following a method in [13]. At each of the 2500 points used for the posterior-predictive method described in Sect. 4.4.5, a Super-Kamiokande fake data set is generated from the MC prediction for that point (using the same reconstructed energy binning as the data fit, defined in Sect. 5.2.2). We then calculate a χ^2 test statistic from the Poisson log-likelihood ratio between the fake data set and the MC prediction, and between the real data set and the MC prediction. If there are ν_i events predicted in each bin and n_i events 'observed' in that bin for either the fake or real data set, then we can define $\vec{\nu}$ and $\vec{n}$ as the total prediction and observation over all bins. From this, the likelihood ratio

$$\lambda = \frac{\mathcal{L}(\vec{n}|\vec{\nu})}{\mathcal{L}(\vec{n}|\vec{n})}$$

enables us to calculate χ^2:

$$\chi^2 = -2\ln\lambda = 2\left[\sum_i n_i \ln\left(\frac{n_i}{\nu_i}\right) + (\vec{\nu} - \vec{n})\right]$$

where the sum runs over all reconstructed energy bins in all four SK samples.

The ND280 samples are not included in the χ^2 calculation in this thesis because the large number of samples and bins would dominate the result, producing a χ^2 that mostly represented how well the ND280 data was fit. The main interest of this thesis is a measurement of neutrino and antineutrino oscillation, for which the most relevant data samples come from the far detector, SK, so an SK-only goodness-of-fit calculation is used. It is, however, still important to understand how well the ND280 data are fit. The *Beam And ND280 Flux extrapolation task Force* (BANFF) group on T2K does a fit to the ND280 data only, with results which are in good agreement with the results presented in this thesis [14]. The goodness-of-fit p-value calculated by the BANFF group from this fit is 0.086 (it was decided in advance to accept any fit with a p-value greater than 0.05).

Once the χ^2 has been obtained, a p-value is calculated as the percentage of samples for which the data fit the MC prediction better than the fake data. Figure 4.10 shows an example distribution of χ^2 between the data and MC plotted against the χ^2 between the fake data and MC, for each of the 2500 samples. The p-value is the proportion of samples for which χ^2(data) $<$ χ^2(fake data), or the proportion of probability in the figure which lies below the black $x = y$ line. As previously discussed, the p-value is interpreted as the probability that we would see this data or something more extreme, assuming that the fit posterior represents the 'true' model of nature. In general, a higher p-value indicates a better fit to the data, but it is up to the analyser or reader to decide what p-value indicates 'acceptable' agreement—common choices would be to accept any fit with a p-value greater than 0.05 or 0.10. However, it is necessary for

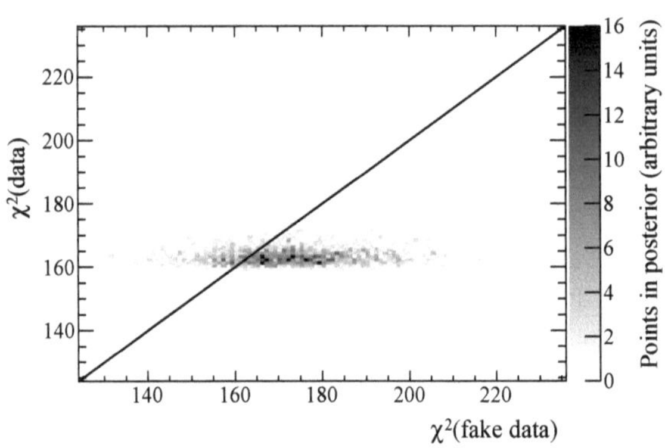

Fig. 4.10 An example distribution of χ^2 between the data and MC prediction (χ^2(data)), plotted against the χ^2 calculated between a fake data set and the MC prediction (χ^2(fake data)), for 2500 samples from the fit posterior, as described in the text. The p-value for the fit is given by the proportion of probability below the *black* $x = y$ line, which in this figure is equal to 0.75

the analyser to define this choice before calculating the p-value to avoid bias once the fit results have been seen. For the results presented in this thesis, a p-value of 0.05 or larger was defined to be 'acceptable'.

References

1. W.R. Gilks, S. Richardson, D.J. Spiegelhalter, *Markov Chain Monte Carlo in Practice* (Chapman & Hall/CRC Interdisciplinary Statistics, Chapman and Hall/CRC, 1995)
2. R.G. Calland, A 3 flavour joint near and far detector neutrino oscillation analysis at T2K, Ph.D. thesis, University of Liverpool (2014), http://www.t2k.org/docs/thesis/059
3. S. Oser, Physics 509C—theory of measurement, slides from lecture series at University of British Columbia, http://www.phas.ubc.ca/oser/p509/
4. C.P. Robert, *The Bayesian Choice: From Decision-Theoretic Foundations to Computational Implementation*, 2nd edn. (Springer, 2007)
5. S. Brooks, A. Gelman, G. Jones, X.-L. Meng (eds.), *Handbook of Markov Chain Monte Carlo*, 1 edn. (Chapman & Hall/CRC Handbooks of Modern Statistical Methods, Chapman and Hall/CRC, 2011). doi:10.1201/b10905
6. N. Metropolis et al., Equation of state calculations by fast computing machines. J. Chem. Phys. **21**(6), 1087–1092 (1953). doi:10.1063/1.1699114
7. W.K. Hastings, Monte carlo sampling methods using markov chains and their applications. Biometrika **57**(1), 97–109 (1970). doi:10.1093/biomet/57.1.97
8. J. Dunkley et al., Fast and reliable markov chain monte carlo technique for cosmological parameter estimation. Month. Not. R. Astron. Soc. **356**(3), 925–936 (2005). doi:10.1111/j.1365-2966.2004.08464.x
9. F. James, MINUIT function minimization and error analysis—reference manual, version 94.1, CERN program library long Writeup D506, https://root.cern.ch/sites/d35c7d8c.web.cern.ch/files/minuit.pdf
10. Peter Mills, Efficient statistical classification of satellite measurements. Int. J. Remote Sens. **32**(21), 6109–6132 (2011). doi:10.1080/01431161.2010.507795
11. K.A. Olive et al. (Particle Data Group), Review of particle physics, Chin. Phys. C **38**, 090001 (2014). (and 2015 update). doi:10.1088/1674-1137/38/9/090001
12. H. Jeffreys, *The Theory of Probability* (OUP Oxford, 1998)
13. A. Gelman, X.L. Meng, H. Stern, Posterior predictive assessment of model fitness via realized discrepancies. Stat. Sin. **6**, 733–759 (1996), http://www3.stat.sinica.edu.tw/statistica/j6n4/j6n41/j6n41.htm
14. M. Hartz et al., Constraining the flux and cross section models with data from the ND280 detector for the 2014/15 oscillation analysis, Technical Report 220, T2K (2015), http://www.t2k.org/docs/technotes/220

Chapter 5
Joint ν+$\bar{\nu}$ Oscillation Analysis: Framework and Validations

The main analysis presented in this thesis is a joint oscillation fit to antineutrino and neutrino data at ND280 and Super-Kamiokande, using the Markov Chain Monte Carlo method. This is described in the next two chapters, both of which are adapted from [1]. This chapter describes the analysis framework: the data samples used from ND280 and SK, the systematic uncertainties considered, and the pre-fit Monte Carlo predictions and comparison to data, as well as some of the validations of the analysis that were performed to verify that the fitter is working as expected. The results of the data fit are then shown in Chap. 6.

To constrain uncertainties due to the neutrino flux and interaction cross sections at SK, data from the ND280 near detector is analysed simultaneously with the SK samples. This effectively considers all T2K data from the near and far detector as one single data set. The motivation to do an analysis in this way comes from the increasing complexity of T2K analyses due to the increasing number of systematic parameters. Other oscillation analyses (including other analyses on T2K) take a different approach: the systematic parameters are constrained using a fit to near detector data, and are then propagated to form inputs for an oscillation analysis using the far detector data. This approach assumes that the expected number of events is linear as a function of these propagated parameters by assuming the log-likelihood behaves quadratically near the maximum—an approximation that is not necessary in the simultaneous-fit approach.

The analysis uses six samples of charged-current (CC) interactions from the ND280 data: FHC (or neutrino mode) ν_μ, RHC (or antineutrino mode) ν_μ, and RHC $\bar{\nu}_\mu$ for each of FGD1 and FGD2 (where the sample is labelled by the sign of the neutrino that must have interacted, inferred from the sign of the muon produced

K.E. Duffy, *First Measurement of Neutrino and Antineutrino Oscillation at T2K*,
Springer Theses, DOI 10.1007/978-3-319-65040-1_5

in the interaction). Both RHC $\bar{\nu}_\mu$ and ν_μ samples are included in order to measure the (relatively large) ν_μ component of the antineutrino-mode beam. The SK detector has no magnetic field and is therefore unable to distinguish between leptons produced in neutrino and antineutrino interactions, so a measurement of the wrong-sign contamination at ND280 is necessary to accurately predict the observables at SK.

Each of the ND280 data samples is split into smaller sub-samples: the FHC samples are divided into three depending on the number of charged pions in the final state ('CC 0π', 'CC 1π', and 'CC Other' sub-samples). Because the statistics in the RHC samples are lower, they are divided into only two sub-samples, by the number of tracks detected: 'CC 1-track' and 'CC N-tracks'. The CC0π and CC 1-track sub-samples give data sets which are dominated by CCQE interactions (the 'signal' at Super-Kamiokande). The CC1π, CC Other, and CC N-tracks sub-samples are dominated by other interactions (largely CC resonant pion production and deep inelastic scattering), and fitting this data allows us to accurately predict the background from other interaction modes at SK. This is the first oscillation analysis to include FGD2 data, which gives the ND280 data the power to constrain cross-section systematics on oxygen.

Four samples of charged-current interactions at SK are included in the fit, labelled FHC 1R$_\mu$, FHC 1R$_e$, RHC 1R$_\mu$, and RHC 1R$_e$.

The data samples and selections are described in Sect. 5.2 and the systematic parameters used to estimate uncertainties due to the neutrino flux, interaction cross sections, and detector response in Sect. 5.3. The pre-fit MC predictions, at both ND280 and SK, are given in Sect. 5.4, as well as the uncertainty on the total number of events in each SK sample due to each group of systematic parameters. Finally, a summary of the validation of the fitter is presented in Sect. 5.5.

5.1 Analysis Strategy

As described in Chap. 4, at each step in the MCMC a posterior probability function is evaluated in order to decide whether or not the step should be accepted. The posterior probability is the product of the likelihood function for the data samples from both detectors and the prior probability functions for the systematic and oscillation parameters. The log of the Bayesian posterior probability for this analysis (which is within a multiplicative constant of the full posterior probability) is:

$$
\begin{aligned}
-\ln(P(\vec{\theta}|D)) = & \sum_{i}^{ND280bins} N_i^{ND,p}(\vec{f},\vec{x},\vec{d}) - N_i^{ND,d} + N_i^{ND,d} ln[N_i^{ND,d}/N_i^{ND,p}(\vec{f},\vec{x},\vec{d})] \\
& + \sum_{i}^{SKbins} N_i^{SK,p}(\vec{f},\vec{x},\vec{skd},\vec{o}) - N_i^{SK,d} + N_i^{SK,d} ln[N_i^{SK,d}/N_i^{SK,p}(\vec{f},\vec{x},\vec{skd},\vec{o})] \\
& + \frac{1}{2}\sum_{i}^{osc}\sum_{j}^{osc} \Delta o_i (V_o^{-1})_{i,j} \Delta o_j \\
& + \frac{1}{2}\sum_{i}^{flux}\sum_{j}^{flux} \Delta f_i (V_f^{-1})_{i,j} \Delta f_j \\
& + \frac{1}{2}\sum_{i}^{xsec}\sum_{j}^{xsec} \Delta x_i (V_x^{-1})_{i,j} \Delta x_j \\
& + \frac{1}{2}\sum_{i}^{nd280det}\sum_{j}^{nd280det} \Delta d_i (V_d^{-1})_{i,j} \Delta d_j \\
& + \frac{1}{2}\sum_{i}^{skdet}\sum_{j}^{skdet} \Delta skd_i (V_{skd}^{-1})_{i,j} \Delta skd_j
\end{aligned}
\tag{5.1}
$$

where the V_{ij} represent covariance matrices constraining those systematic parameters which do not have flat priors. These are labelled f for flux, x for cross section, d for ND280 detector, skd for SK detector, and o for oscillation. A separate covariance matrix is used for each group of parameters, indicating that no prior correlations are considered between different types of parameters (for example, prior correlations between different flux bin-normalisation parameters are considered, but not between flux and cross-section parameters). It is a slight shortcut to write the oscillation priors in this way: as described below, there are non-flat priors only on the parameters $\sin^2\theta_{13}$, $\sin^2\theta_{12}$, and Δm^2_{21}, and no correlations between oscillation parameters in the prior. The diagonal terms $(V_o)_{i,i}$ representing $\sin^2\theta_{23}$, Δm^2_{32}, and δ_{CP} are zero, resulting in flat priors on those oscillation parameters. $N_i^{ND,p}$ and $N_i^{SK,p}$ represent the number of predicted events (from the MC) in a particular bin i for ND280 or SK respectively, for some given values of the model parameters $\vec{f}$, $\vec{x}$, $\vec{d}$, $\vec{skd}$, and $\vec{o}$, and $N_i^{ND,d}$ and $N_i^{SK,d}$ the number of data events.

The binning of the data events in this analysis is in 2D lepton p–$\cos\theta$ space for the ND280 data and 1D reconstructed neutrino or antineutrino energy space for the four SK samples. The reconstructed neutrino energy at SK is determined from the kinematics of the measured lepton, assuming the event was a CCQE interaction on a stationary bound nucleon. The reconstructed energy formula for a CCQE muon neutrino interaction with a neutron was given in Eq. 3.1.

One set of oscillation parameters (according to the PMNS model) is used to describe all neutrino and antineutrino data, with all six oscillation parameters allowed to float in the fit. We choose the priors for the oscillation parameters $\sin^2\theta_{23}$, $|\Delta m^2_{32}|$, and δ_{CP} to be flat. The prior for $\sin^2\theta_{13}$ may be flat (described as 'without reactor constraint') or may be Gaussian according to the measurement of $\sin^2 2\theta_{13}$ reported from reactor experiments in the 2015 PDG [2]: $\sin^2 2\theta_{13} = 0.085 \pm 0.005$ (referred to as 'with reactor constraint'). The fit will be repeated twice—once with the reactor constraint and once without. Gaussian priors from the PDG are always used for the solar parameters, $\sin^2\theta_{12}$ and Δm^2_{12}, to which T2K has little sensitivity: $\sin^2 2\theta_{12} = 0.846 \pm 0.021$ and $\Delta m^2_{21} = (7.53 \pm 0.18) \times 10^{-5}\text{eV}^2/\text{c}^4$ [2]. Additionally, we use a prior for the mass hierarchy (encoded as the sign of Δm^2_{32}) such that the probability of the normal and inverted hierarchies are equal, i.e. P(NH)=P(IH)=0.5. When the Markov chain runs there is a 50% probability for the proposal function to flip between normal and inverted hierarchy as it proposes each step.

5.2 Event Selection

5.2.1 ND280

The ND280 selections for this analysis include both neutrino- and antineutrino-mode data with interactions in FGD1 or FGD2. The neutrino-mode data was collected between November 2010 and February 2015 (T2K runs 2–4, 5, and 6), corresponding to an exposure of 5.82×10^{20} POT, and the antineutrino mode data was collected between June 2014 and April 2015 (T2K runs 5 and 6), with an exposure of 2.84×10^{20} POT.

In both neutrino-mode and antineutrino-mode data we wish to select only charged-current events, in which an interaction occurs in FGD1 or FGD2 and the interaction products are forward-going and measured in at least one TPC. Beyond this broad selection, the neutrino-mode data selection is further split into six categories: ν_μ CC 0π (dominated by CCQE interactions), ν_μ CC 1π (dominated by CC RES interactions), and ν_μ CC Other (dominated by CC DIS interactions) for both FGD1 and FGD2, where the interaction categories are defined in Sect. 3.4. The antineutrino mode data selection is split into a total of eight categories: ν_μ CC1-track and $\bar{\nu}_\mu$ CC 1-track (dominated by CCQE interactions), and ν_μ CC N-track and $\bar{\nu}_\mu$ CC N-track (with N>1; a mixture of CC RES and DIS interactions), again for both FGD1 and FGD2.

The ND280 data selections are described in detail in [3–7], and summarised below.

5.2.1.1 ND280 ν_μ Selection

The aim of the ND280 ν_μ event selection is to define a sample of neutrino interactions which originate in the FGD1 or FGD2 detector and contain at least one muon track with negative charge crossing the TPC downstream from the FGD in which the interaction occurred. The selections for the ν_μ samples in neutrino and antineutrino mode are:

Event Quality The event must occur in a defined proton bunch from a beam spill in which the data quality was deemed to be good.

Total Multiplicity In the selected event, at least one track must cross one of the TPCs.

Fiducial Volume The highest momentum negative-charge track in the event must have its origin in the FGD1 or FGD2 fiducial volume.

TPC Measurement The TPC reconstruction is less reliable for short tracks, so we require that the highest-momentum negative-charge track in the event must cross more than 18 layers in a TPC, where layer orientation is defined by the local track angle.

TPC PID The particle identification (PID) of the highest momentum negative-charge track must be consistent with a muon.

Highest Momentum Track The highest momentum *negative-charge* track must be the highest momentum track in the event. This is selected as the μ^- candidate.

Upstream Background Veto Reconstruction failures can make a muon track falsely appear to start in the FGD (for example, a muon from an interaction in the P0D which undergoes a large scatter in FGD1 can be reconstructed as two separate tracks). To exclude these events, we impose a veto on backwards-going events or events coming from the P0D or Magnet. For the same reason, in the FGD2 selection the event is vetoed if the second-highest momentum track starts in the FGD1 fiducial volume.

External FGD If the event has at least one reconstructed FGD-only track, it must not have its starting position in the two layers of the FGD closest to the upstream edge.

After these steps, the selection is split into the CC 0π, CC 1π, and CC Other (for neutrino mode) or CC 1-track and CC N-track (for antineutrino mode) samples.

The presence of a charged pion can be inferred either from a pion track in a TPC or FGD or from the delayed 'Michel' electron produced in pion decay, which will occur outside the beam time window. The neutrino mode CC 0π sample is defined by rejecting events with a $\pi^\pm$ or $e^\pm$ in a TPC, a $\pi^\pm$ in an FGD, or a Michel electron in an FGD. The CC 1π sample is defined by rejecting events with a π^- or $e^\pm$ in the TPC, and selecting events in which the sum of the number of π^+ found in the TPC and the number of Michel electrons measured is equal to 1. If no Michel electrons are measured, the sum of the number of π^+ measured in the TPC and the number of π^+ measured in the FGD must be equal to 1. The CC Other sample contains all events that pass the selection cuts above but do not fall into the CC 0π or CC 1π samples.

In antineutrino mode, the CC 1-track sample is defined as having only one reconstructed track, the highest momentum negative-charge track. The CC N-track sample contains all other events.

5.2.1.2 ND280 $\bar{\nu}_\mu$ Selection

The aim of the ND280 $\bar{\nu}_\mu$ event selection is to define a sample of antineutrino interactions which originate in the FGD1 or FGD2 detector and contain at least one muon track with positive charge crossing the TPC downstream from the FGD in which the interaction occurred. The selections for the $\bar{\nu}_\mu$ samples in neutrino and antineutrino mode are:

Event Quality The event must occur in a defined proton bunch from a beam spill in which the data quality was deemed to be good.

Total Multiplicity In the selected event, at least one track must cross one of the TPCs.

Fiducial Volume The highest momentum positive-charge track in the event must have its origin in the FGD1 or FGD2 fiducial volume.

TPC Measurement The TPC reconstruction is less reliable for short tracks, so we require that the highest-momentum positive-charge track in the event must must cross more than 18 layers in a TPC, where layer orientation is defined by the local track angle.

TPC PID The PID of the highest momentum positive-charge track must be consistent with a muon.

Highest Momentum Track The highest momentum *positive-charge* track must be the highest momentum track in the event. This is selected as the μ^+ candidate.

Upstream Backgroud Veto Reconstruction failures can make a muon track falsely appear to start in the FGD (for example, a muon from an interaction in the P0D which undergoes a large scatter in FGD1 can be reconstructed as two separate tracks). To exclude these events, we impose a veto on backwards-going events or events coming from the P0D or Magnet. For the same reason, in the FGD2 selection the event is vetoed if the second-highest momentum track starts in the FGD1 fiducial volume.

External FGD If the event has at least one reconstructed FGD-only track, it must not have its starting position in the two layers of the FGD closest to the upstream edge.

After these steps, the selection is split into the CC 1-track and CC N-track samples. The CC 1-track sample is defined as having only one reconstructed track, the highest momentum positive-charge track. The CC N-track sample contains all other events.

The number of data events in each sample are shown in Table 5.1. All samples are binned in a 2D p–$\cos\theta$ grid with bin definitions as follows:

- CC 0π and CC Other bin edges:
  p (MeV/c): 0, 300, 400, 500, 600, 700, 800, 900, 1000, 1250, 1500, 2000, 3000, 5000, 30000
  $\cos\theta$: -1.0, 0.6, 0.7, 0.8, 0.85, 0.9, 0.92, 0.94, 0.96, 0.98, 0.99, 1
- CC 1π bin edges:
  p (MeV/c): 0, 300, 400, 500, 600, 700, 800, 900, 1000, 1250, 1500, 2000, 5000, 30000
  $\cos\theta$: -1.0, 0.6, 0.7, 0.8, 0.85, 0.9, 0.92, 0.94, 0.96, 0.98, 0.99, 1
- CC 1-Track bin edges:
  p (MeV/c): 0, 500, 900, 1200, 2000, 10000
  $\cos\theta$: -1.0, 0.8, 0.92, 0.98, 1

Table 5.1 Number of events in each of the ND280 data samples

Data sample		POT	Number of events in ND280 data
FHC	FGD1 ν_μ CC 0π	5.82×10^{20}	17354
	FGD1 ν_μ CC 1π		3984
	FGD1 ν_μ CC Other		4220
RHC	FGD1 $\bar{\nu}_\mu$ CC 1-Track	2.84×10^{20}	2663
	FGD1 $\bar{\nu}_\mu$ CC N-Track		775
	FGD1 ν_μ CC 1-Track		989
	FGD1 ν_μ CC N-Track		1001
FHC	FGD2 ν_μ CC 0π	5.82×10^{20}	17650
	FGD2 ν_μ CC 1π		3383
	FGD2 ν_μ CC Other		4118
RHC	FGD2 $\bar{\nu}_\mu$ CC 1-Track	2.84×10^{20}	2762
	FGD2 $\bar{\nu}_\mu$ CC N-Track		737
	FGD2 ν_μ CC 1-Track		980
	FGD2 ν_μ CC N-Track		936

- CC N-Track bin edges:
  p (MeV/c): 0, 600, 1000, 1500, 2200, 10000
  $\cos\theta$: -1.0, 0.8, 0.9, 0.97, 1

The momentum distributions of each of the data samples are shown graphically compared to the MC prediction in Figs. 5.5, 5.6, 5.7, and 5.8, in Sect. 5.4.4. The $\cos\theta$ distributions are not included because of space limitations, but can be found in [8].

5.2.2 Super-Kamiokande

This analysis uses all the T2K data collected at SK in runs 1–7b, between January 2010 and May 2016. This corresponds to a total POT of 7.002×10^{20} in neutrino mode and 7.471×10^{20} POT in antineutrino mode. The data and MC are separated into four selections: single-ring muon-like ($1R_\mu$), and single-ring electron-like ($1R_e$), for both neutrino-mode and antineutrino-mode beam.

There is no difference in the event selections (which are described in detail in [9] and summarised here) between the forward and reverse horn current samples. For the $1R_\mu$ samples, the selection is as follows:

Fiducial Volume Event must be fully contained in the fiducial volume at Super-Kamiokande.

Cherenkov Rings Only events for which exactly one ring is found by the ring counting algorithm are accepted.

PID The ring must be identified as muon-like by the PID algorithm.

Reconstructed Momentum The reconstructed momentum of the muon must be greater than 200 MeV/c.

Decay Electrons The number of decay electrons must be less than or equal to one.

The selected events in both the forward and reverse horn curent 1R$_\mu$ samples are binned in reconstructed energy, in 50 MeV-width bins from 0–3 GeV, then in bins with edges at 3.25, 3.5, 3.75, 4.0, 4.5, 5.0, 5.5, 6.0, 7.0, 8.0, 9.0, 10.0, and 30.0 GeV.

For the 1R$_e$ samples, selected events must pass the criteria:

Fiducial Volume Event must be fully contained in the fiducial volume at Super-Kamiokande.

Cherenkov Rings Only events for which exactly one ring is found by the ring counting algorithm are accepted.

PID The ring must be identified as electron-like by the PID algorithm.

Visible Energy The amount of Cherenkov light measured in the detector must be greater than that of a 100 MeV electromagnetic shower.

Decay Electrons Events with any decay electrons are rejected.

Reconstructed Energy The reconstructed energy of the neutrino must be smaller than 1250 MeV.

π^0 **Rejection** The event must pass a π^0 rejection cut: $\ln(\mathcal{L}_{\pi^0}/\mathcal{L}_e) < 175 - 0.875 \times m_{\gamma\gamma}$.

The π^0 rejection cut [10] is necessary because of the similarity between Cherenkov rings produced by photons and electrons—if the two photon rings overlap or one ring is not detected, a π^0 in the SK detector could be misreconstructed as an electron. Each event is reconstructed under the π^0 hypothesis—assuming that the observed signal at SK is actually two rings produced by the two decay photons from a π^0—and the reconstructed invariant mass of these two photons, $m_{\gamma\gamma}$, is calculated. The likelihood of this hypothesis is calculated and compared to the likelihood under the e-like hypothesis, in which it is assumed that there is only one ring which was produced by an electron. The cut itself is then defined as a line in $\ln(\mathcal{L}_{\pi^0}/\mathcal{L}_e)$ vs $m_{\gamma\gamma}$ space, as shown in Fig. 5.1. The MC prediction for oscillated ν_e CCQE interactions and NCπ^0 interactions in $\ln(\mathcal{L}_{\pi^0}/\mathcal{L}_e)$ vs $m_{\gamma\gamma}$ space is shown, along with the diagonal line that defines the π^0 rejection criterion—events above the line are classified as more π^0-like than e-like and are rejected from the selection.

The data for the forward and reverse horn current 1R$_e$ samples are binned in reconstructed energy, in 50 MeV-width bins from 0–1.25 GeV.

Table 5.2 summarises the number of data events in each SK sample. The data are shown graphically compared to the MC prediction in Fig. 5.9 in Sect. 5.4.5.

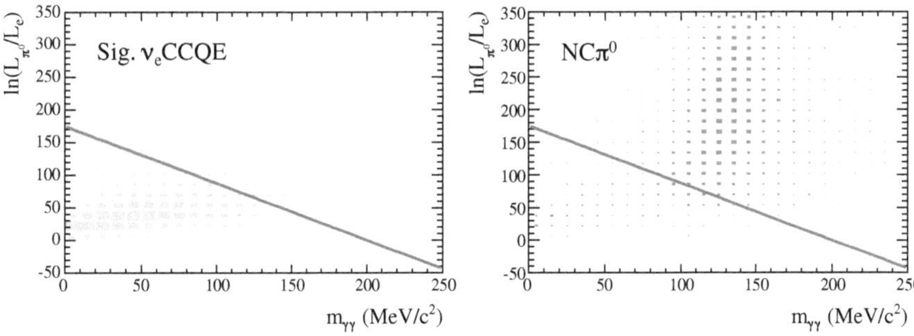

Fig. 5.1 The 2D MC prediction for oscillated ν_e CCQE interactions (*left*) and NCπ^0 interactions (*right*) in $\ln(\mathcal{L}_{\pi^0}/\mathcal{L}_e)$ vs the reconstructed invariant mass $m_{\gamma\gamma}$. The diagonal line indicates the π^0 rejection criterion, and events lying above the line are rejected as π^0 background. The size of each box is proportional to the number of events in the bin. Both figures use the same scale for representing the number of events and are normalised to the same POT. Figure from [10]

Table 5.2 Number of events in each of the SK data samples

Data sample	POT	Number of events in SK data
FHC $1R_\mu$	7.002×10^{20}	125
FHC $1R_e$		32
RHC $1R_\mu$	7.471×10^{20}	66
RHC $1R_e$		4

5.3 Systematic Uncertainties

5.3.1 Flux

The T2K beam simulation was described in Sect. 3.1.3. Uncertainty in the neutrino flux prediction comes from uncertainties in the hadron production model, out-of-target material modelling, beam conditions, alignment of beam line components, and horn field asymmetry. The uncertainty near the flux peak for the right-sign flux (ν in neutrino mode or $\bar{\nu}$ in antineutrino mode) is around 8%, and the dominant uncertainty comes from the hadron production model. At low energy this is largely due to uncertainties on the pion production cross section and multiplicity, and secondary nucleon production. At high energies, the uncertainty on kaon production dominates.

100 systematic parameters are used to describe the neutrino flux uncertainty. All parameters represent bin-content normalisations in bins of neutrino energy, and span eight categories: FHC ν_μ (11 bins), FHC $\bar{\nu}_\mu$ (5 bins), FHC ν_e (7 bins), FHC $\bar{\nu}_e$ (2 bins), RHC ν_μ (5 bins), RHC $\bar{\nu}_\mu$ (11 bins), RHC ν_e (2 bins), and RHC $\bar{\nu}_e$ (7 bins). This gives 50 bins, which are repeated twice—once for the flux uncertainty at ND280 and once for the uncertainty at SK—resulting in 100 parameters in total. Within each

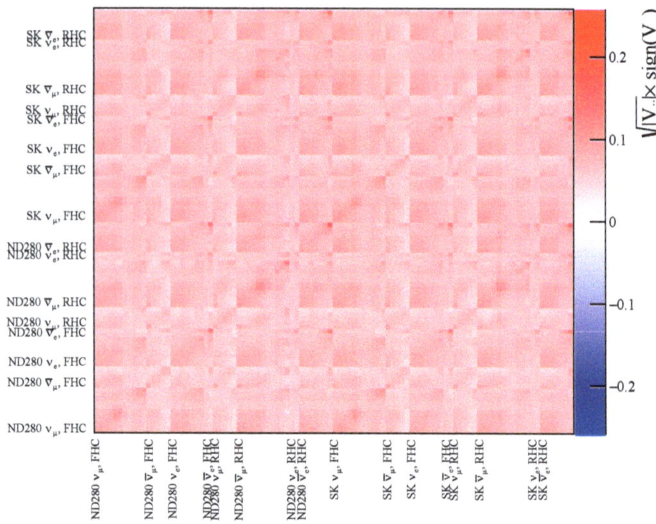

Fig. 5.2 The fractional covariance matrix used for the flux systematic parameters. The bin labels show where each group of bins (as given in the text) begins. Within each sample, the bins iterate over true neutrino energy

sample, the flux uncertainties are split into bins of true neutrino energy, with finer binning around the oscillation maximum ($\sim$0.6 GeV). A fractional covariance matrix for the flux systematic uncertainty [11, 12] is provided by the beam group on T2K, as shown in Fig. 5.2.

All flux systematic parameters have a Gaussian prior with width equal to the standard deviation from the fractional covariance matrix, and a mean value of 1. These are applied as a normalisation: the number of events in a given bin is scaled according to the value of the corresponding systematic parameter by applying weights to MC events on an event-by-event basis.

5.3.2 *Cross Section*

The MC used for the predictions at ND280 and SK is based on a Relativistic Fermi Gas with relativistic Random Phase Approximation correction to the W propagator (RFG + rel.RPA) model for charged-current interactions. The parameterisation and uncertainties were calculated by the *Neutrino Interaction Working Group* (NIWG) on T2K, using information from fits to external data from the MINERνA and MiniBooNE experiments [13] (for the CCQE parameters), and from the ANL and BNL experiments [14] (for the parameters governing resonant pion production). Further details about these external data fits and subsequent recommendations for the cross-section model can be found in [15–17].

There are two types of cross-section systematic parameter, termed 'shape' and 'normalisation', which are treated differently. Normalisation parameters have the same effect on all events, regardless of energy. For these parameters, each event is simply weighted by the value of the parameter.

Shape parameters, on the other hand, may affect events with different energies (or different momentum transfer) differently. In this case it is not appropriate to weight all events by the same value—the value by which an event must be weighted to account for (for example) a $+3\sigma$ shift in a shape parameter depends on its kinematics. Instead, response functions are created by evaluating the changed weight for a particular MC event at evenly spaced points in the parameter value (between 2 and 13 points, depending on the parameter) and creating a cubic spline to interpolate between the points. This spline is then evaluated at the desired value of the parameter, and the result applied to the event as a weight. For the ND280 MC, an individual spline is created for each MC event. For the SK MC, events are binned in reconstructed and true neutrino energy, and splines created from the averaged weight calculated for a particular kinematic bin. This was done to reduce the computational load on the analysis, due to the large amounts of SK MC; each MC event is then weighted according to the evaluated spline for the kinematic bin into which it falls.

The following systematic parameters (listed in Table 5.3) are considered to calculate the total cross-section uncertainty:

M_A^{QE} (shape parameter) The axial mass, which appears in the axial form factor for CCQE scattering of neutrinos on nucleons. This is an effective parameter: it represents the axial mass M_A, but only when it is applied to CCQE interactions. A second effective parameter is used for M_A in CC RES interactions. Because of tension between CCQE data sets from MINERνA, MiniBooNE, and T2K, the NIWG group recommended using a flat prior for this parameter, with no correlations with any other parameters.

p_F (shape parameter) In the RFG model, the nuclear potential is determined entirely by two parameters: p_F, the average Fermi momentum of a nucleon, and E_B, the average binding energy of a nucleon. The parameters vary by nuclear target, so different parameters are used for interactions on carbon and oxygen (because we measure interactions on scintillator—made from hydrocarbons—and water. These parameters are not needed to describe interactions on hydrogen because the hydrogen nucleus is made from a single proton). Because of tension in the CCQE data from MINERνA, MiniBooNE, and T2K, the NIWG group also recommended using flat priors for these parameters, with no correlations with any other parameters. The p_F parameters are valid only for a reduced parameter range, given in Table 5.3. While this is inconsistent with the use of flat priors, it was necessary for practical reasons—the reweighting software used for this analysis is not able to produce accurate results outside this range (something that will be addressed in future analyses).

E_B (shape parameter) The average binding energy of a nucleon in the RFG model. As for p_F, the parameter varies by nuclear target so two parameters are used: one which applies to interactions on carbon, and one which applies to interactions on oxygen. The prior mean and uncertainty are constrained by fits to external data from MINERνA and MiniBooNE. The E_B parameters are valid over a reduced parameter range, given in Table 5.3.

2p-2h (normalisation parameter) Parameter that adjusts the number of 2p-2h events (in which the neutrino interaction produces multiple nucleons but no pions, so these events are included in the CC 0π/CC 1-track and single-ring selections as a background for true CCQE events). The way to correctly apply this normalisation to events on carbon and oxygen was carefully considered, because there is little experimental data to inform how the cross section scales with target nucleus. In the end a conservative approach was chosen: two separate parameters to describe interactions on carbon (2p-2h ^{12}C) and oxygen (2p-2h ^{16}O) with flat, uncorrelated priors.

$\bar{\nu}$ **2p-2h (normalisation parameter)** There are two major models for 2p-2h interactions: the model by Nieves et. al. [18, 19] which is implemented in the T2K simulation, and a model by Martini et. al. [20]. The difference between the predictions of these two models is different for neutrino and antineutrino interactions, which provides motivation for an extra normalisation factor to take into account these potential differences. The purpose of this extra parameter is to allow the normalisation of the 2p-2h distributions for neutrinos and antineutrinos to differ (as it is a normalisation factor, it does not affect the shape of the distributions). Antineutrino 2p-2h interactions should be weighted by the product of either the 2p-2h ^{12}C parameter and the 2p-2h $\bar{\nu}$ parameter, or by the product of the 2p-2h ^{16}O parameter and the 2p-2h $\bar{\nu}$ parameter, depending on the nucleus on which the interaction took place. The $\bar{\nu}$ 2p-2h normalisation parameter has a flat prior, which is uncorrelated with any other parameters.

$\mathbf{C_5^{ARES}}$ **(shape parameter)** The normalisation of the axial form factor for resonant pion production in neutrino interactions on nucleons. The prior uncertainty and mean are constrained by fits to external hydrogen and deuterium scattering data from the ANL and BNL bubble chamber experiments.

$\mathbf{M_A^{RES}}$ **(shape parameter)** The axial mass which appears in the axial form factor for resonant pion production in neutrino interactions on nucleons. As for M_A^{QE}, M_A^{RES} is an effective parameter: it represents the axial mass M_A, but only when it is applied to CC RES interactions. The prior uncertainty and mean are constrained by fits to external hydrogen and deuterium scattering data from the ANL and BNL bubble chamber experiments.

Isospin=$\frac{1}{2}$ Background (shape parameter) Size of the isospin=$\frac{1}{2}$ non-resonant background to isospin=$\frac{3}{2}$ resonant pion production interactions. The prior uncertainty and mean are constrained by fits to external hydrogen and deuterium scattering data from the ANL and BNL bubble chamber experiments.

CC ν_e / ν_μ and CC $\bar{\nu}_e/\bar{\nu}_\mu$ (normalisation parameters) Parameters that account for the difference in ν_e and ν_μ (and $\bar{\nu}_e$ and $\bar{\nu}_\mu$) cross-sections. It is necessary to include this uncertainty because only ν_μ and $\bar{\nu}_\mu$ interactions are measured at ND280, and the results of fitting the ND280 data are used to predict ν_e and $\bar{\nu}_e$ event rates at SK. Two parameters are used: CC ν_e/ν_μ, which applies to CC

ν_e interactions only, and CC $\bar{\nu}_e/\bar{\nu}_\mu$, which applies to CC $\bar{\nu}_e$ interactions only. The covariance matrix for these two parameters includes an uncorrelated contribution to account for radiative corrections and an anti-correlated contribution to allow for second-class currents as calculated in [21]:

$$V_{\nu_e,\bar{\nu}_e} = V_{RadCorr} + V_{SCC} = \begin{bmatrix} 0.02^2 & 0.0 \\ 0.0 & 0.02^2 \end{bmatrix} + \begin{bmatrix} 0.02^2 & -0.02^2 \\ -0.02^2 & 0.02^2 \end{bmatrix}$$
$$= \begin{bmatrix} 2 \times 0.02^2 & -0.02^2 \\ -0.02^2 & 2 \times 0.02^2 \end{bmatrix}$$

CC Other (shape parameter) Energy-dependent uncertainty applied to interactions categorised as 'CC Other'. These include CC multi-π production, CC DIS interactions, and CC non-pion resonant production (which, in our MC, includes γ, η, and K production). This parameter is not directly constrained by external data, but from external data sets it is known that the uncertainty is around 10% at $E_\nu = 4$ GeV. The uncertainty assigned is:

$$\delta_{\sigma_{CC Other}} = \frac{0.4}{E_\nu}$$

This approaches infinity as $E_\nu \to 0$, but the production threshold of these events is 0.6 GeV.

CC Coherent (normalisation parameter) Parameter that adjusts the number of CC coherent pion production interactions, where 'coherent' indicates that the neutrino interacts with the entire nucleus, not its constituent nucleons. This parameter is not directly constrained by external data, but the 30% uncertainty assigned to it is informed by comparisons to MINERνA data [22]. Two parameters are used, which are 100% correlated in the prior: one which is applied to interactions on ^{12}C and one which is applied to interactions on ^{16}O.

NC Coherent (normalisation parameter) Parameter that adjusts the number of NC coherent pion production interactions. This parameter is not directly constrained by external data, but the 30% uncertainty assigned to it is motivated by a 15% discrepancy between NEUT and an external data set known to have a 20% systematic uncertainty.

NC 1γ (normalisation parameter) Parameter that adjusts the number of NC non-pion resonant interactions which produce a γ. There is no control sample in ND280 or from external data which provides a limit on this normalisation, so a 100% uncertainty is used.

NC Other (normalisation parameter) Parameter that adjusts the number of 'NC Other' events. These include NC multi-π production, NC DIS interactions, and NC non-pion resonant interactions which produce a η or K (but not NC non-pion resonant interactions which produce a γ). Unlike the 'CC Other' parameter, this is applied as a normalisation with a 30% uncertainty rather than an energy-dependent parameter due to poor constraints from external data for NC Other interactions. Two separate, uncorrelated, parameters are used to describe the NC

Other normalisation at the near and far detectors: NC Other (near), which applies at ND280 only, and NC Other (far), which applies at SK only. The components that make up the NC Other category at SK and ND280 are not the same, so applying one parameter to all of them—especially if that parameter is constrained at ND280—is not appropriate.

FSI parameters (ND280 only, shape parameters) Final State Interactions are interactions of hadrons produced by neutrino interactions before they escape the nucleus. The effect of FSI is to migrate events between different observable topologies in the detector and change the particle kinematics of hadrons that make it out of the nucleus. For example, a Δ decay event that produces a single pion should be categorised as a CC 1π or CC N-track event. However, if the pion interacts and is absorbed in the nucleus, no pion will be seen in the final state and the event will be selected as CC 0π or CC 1-track. FSI can include QE scattering, particle absorption, particle production, and charge exchange, and uncertainties due to FSI are estimated from fits of the microscopic cascade model parameters in NEUT to pion scattering on carbon. Six parameters are used to describe the different elements of FSI uncertainties: one parameter each for pion production, pion absorption, low and high energy charge exchange, and low and high energy inelastic interactions. These are treated as factorisable.

Both neutrinos and antineutrinos are simulated using the same model, with uncertainties according to the same model parameters (with the exception of the $\bar{\nu}$ 2p-2h and CC $\bar{\nu}_e/\bar{\nu}_\mu$ parameters, which apply to antineutrinos only) to ensure that the model is consistent for both. Some parameters should be applied to the near and far detectors separately, and some should apply to both. The NC 1γ, CC ν_e/ν_μ, and CC $\bar{\nu}_e/\bar{\nu}_\mu$ parameters should apply only at the far detector (because there are very few data events of these types in the ND280 selections which could be used to constrain the parameters), as well as the NC Other (far) parameter. The Final State Interaction (FSI) parameters apply at ND280 only: uncertainties due to FSI at SK are incorporated with the SK detector uncertainties as described in Sect. 5.3.4, in order to reduce the total number of parameters in the analysis.

5.3.3 ND280 Detector

The ND280 detector systematics are described by 580 bin-content normalisation parameters, each of which has nominal value equal to 1. These are discussed in more detail in [3, 8, 12, 23]. A fractional covariance matrix was provided by the *Beam And ND280 Flux extrapolation task Force* (BANFF) group on T2K, which was evaluated by producing 2000 instances of the inputs for the detector systematics, with systematic and statistical fluctuations, and running the full detector systematic analysis for each instance. This allows for event weights and observables to vary, and for events to move in and out of samples as well as migrate between bins within a sample, while also accounting for the MC statistical uncertainty.

Table 5.3 Cross section parameters used for this analysis, showing the valid range of the parameter, prior mean, nominal value in NEUT, and prior uncertainty. The type of parameter (shape or normalization) is also given

Parameter	Valid Range	Prior Mean	Nominal	1σ Uncertainty	Class
M_A^{QE} (GeV/c^2)	all	flat prior	1.20	flat prior	shape
p_F ^{12}C (MeV/c)	200–275	flat prior	217	flat prior	shape
2p-2h ^{12}C	all	flat prior	1.0	flat prior	norm
E_B ^{12}C (MeV)	12–42	25	25	9	shape
p_F ^{16}O (MeV/c)	200–275	flat prior	225	flat prior	shape
2p-2h ^{16}O	all	flat prior	1.0	flat prior	norm
E_B ^{16}O (MeV)	12–42	27	27	9	shape
2p-2h $\bar{\nu}$	all	flat prior	1.0	flat prior	norm
C_5^{ARES}	all	1.01	1.01	0.12	shape
M_A^{RES} (GeV/c^2)	all	0.95	0.95	0.15	shape
Isospin $= \frac{1}{2}$ Background	all	1.30	1.30	0.20	shape
CC ν_e/ν_μ	all	1.0	1.0	0.04	norm
CC $\bar{\nu}_e/\bar{\nu}_\mu$	all	1.0	1.0	0.04	norm
CC Other	all	0.0	0.0	0.40	shape
CC Coh ^{12}C	all	1.0	1.0	0.30	norm
CC Coh ^{16}O	all	1.0	1.0	0.30	norm
NC Coh	all	1.0	1.0	0.30	norm
NC 1γ	all	1.0	1.0	1.00	norm
NC Other (near)	all	1.0	1.0	0.30	norm
NC Other (far)	all	1.0	1.0	0.30	norm
FSI Inel. Low E (ND280)	all	0.0	0.0	0.41	shape
FSI Inel. High E (ND280)	all	0.0	0.0	0.34	shape
FSI Pion Prod. (ND280)	all	0.0	0.0	0.50	shape
FSI Pion Abs. (ND280)	all	0.0	0.0	0.41	shape
FSI Ch. Exch. Low E (ND280)	all	0.0	0.0	0.57	shape
FSI Ch. Exch. High E (ND280)	all	0.0	0.0	0.28	shape

Systematic correlations between FGD1 and FGD2 must also be correctly accounted for. This is done by generating two sets of detector systematic parameters for each instance, with one set of variations applied to FGD1-selected events and the other set applied to FGD2-selected events.

There are 580 ND280 detector bin-normalisation parameters in total: 70 each for the ν_μ CC 0π, ν_μ CC 1π, and ν_μ CC Other samples in both FGD1 and FGD2 (420 bins in total for the neutrino-mode samples), and 20 each for the CC 1-track and CC N-track samples for ν_μ and $\bar{\nu}_\mu$ in both FGD1 and FGD2 (160 bins in total for the antineutrino-mode samples). Within each sample, the parameters represent

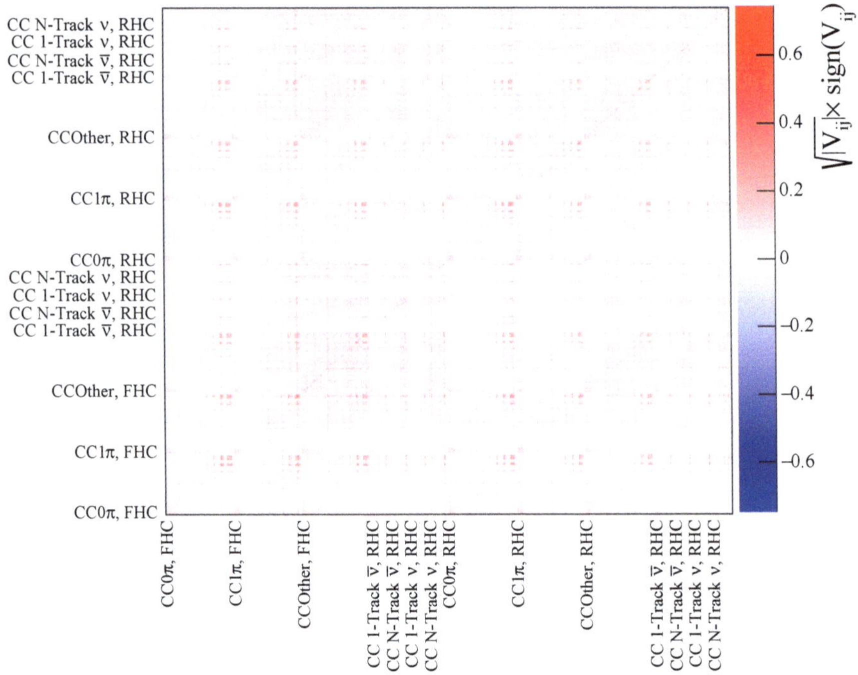

Fig. 5.3 The fractional covariance matrix used for the ND280 detector systematic parameters. The bin labels show where each group of bins (as given in the text) begins. Within each ND280 sub-sample, the bins iterate over $\cos\theta$ from low to high for the lowest momentum bin, then from low to high for the second-lowest momentum bin, and so on

different bins in lepton momentum and angle (p and $\cos\theta$). These iterate over $\cos\theta$ bins from low to high for the lowest momentum bin, then from low to high for the second-lowest momentum bin, and so on.

The binning in p–$\cos\theta$ for the antineutrino-mode detector uncertainties is the same as the binning used for the fit. For the neutrino-mode samples, a coarser binning was chosen than the binning used for fitting the data in order to reduce the number of parameters used in the fit. This does not result in a significant loss of accuracy because the size of the detector systematic uncertainties is typically smaller than the size of the flux and cross section uncertainties.

The fractional covariance matrix is shown in Fig. 5.3. Similarly to the flux uncertainties, changes in the ND280 detector systematic parameters are applied as a normalisation: each event is weighted according to the value in the bin corresponding to the event's reconstructed momentum and angle.

5.3.4 Super-Kamiokande Detector

The Super-Kamiokande detector uncertainties are described by 37 bin-content normalisation parameters, correlated between the ν mode and $\bar{\nu}$ mode $1R_e$ and $1R_\mu$ samples. The correlations and uncertainties on these parameters are provided to analysers in the form of a fractional covariance matrix by the T2K-SK group [24]. To produce this matrix, the uncertainties on the topological selection cuts described in Sect. 5.2.2 are estimated using atmospheric neutrino and cosmic ray data at SK, and are then combined and converted into an uncertainty in reconstructed energy space using toy MC.

The fractional covariance matrix consists of 37 bin-content normalisation parameters, each representing a reconstructed energy bin, over 17 categories. The first eight categories are repeated twice, once for the FHC samples and once for the RHC samples: $1R_e$ oscillated ν_e+$\bar{\nu}_e$ (3 bins), $1R_e$ ν_μ+$\bar{\nu}_\mu$ (3 bins), $1R_e$ beam ν_e+$\bar{\nu}_e$ (3 bins), $1R_e$ NC (3 bins), $1R_\mu$ ν_μ+$\bar{\nu}_\mu$ CCQE (including CC 2p-2h) (3 bins), $1R_\mu$ ν_μ+$\bar{\nu}_\mu$ CCnonQE (1 bin), $1R_\mu$ ν_e+$\bar{\nu}_e$ (1 bin), and $1R_\mu$ NC (1 bin).

The final parameter in the fractional covariance matrix is the energy scale uncertainty. This is currently the only systematic uncertainty which is not applied as a weight to an MC event; instead, the reconstructed energy itself is adjusted according to the value of the parameter, before the MC is binned. The energy scale uncertainty used for this analysis is 2.4% regardless of E_ν.

The final fractional covariance matrix used in the oscillation fit does not only contain the SK detector uncertainties. It also contains systematic uncertainties from hadron re-interactions in the target nucleus (Final State Interaction, or FSI uncertainties) and in the detector (Secondary Interaction, or SI uncertainties), as well as systematic uncertainties due to photo-nuclear (PN) interactions. PN interactions only affect the $1R_e$ sample; they allow for the absorption of photons without subsequent re-emission above the Cherenkov threshold, which can cause pion interactions to be mistaken as ν_e, and are assigned a 100% uncertainty. The FSI+SI and PN uncertainties were calculated by re-evaluating the MC for different amounts of FSI, SI and PN, as described in [25], and made into fractional covariance matrices in the same binning as used for the detector systematics. One matrix was made for the FSI+SI uncertainties, and a second for the $1R_e$ sample PN uncertainties. These two matrices were then combined linearly with the detector-only matrix—because the uncertainties are uncorrelated—to produce the final covariance matrix, which is shown in Fig. 5.4.

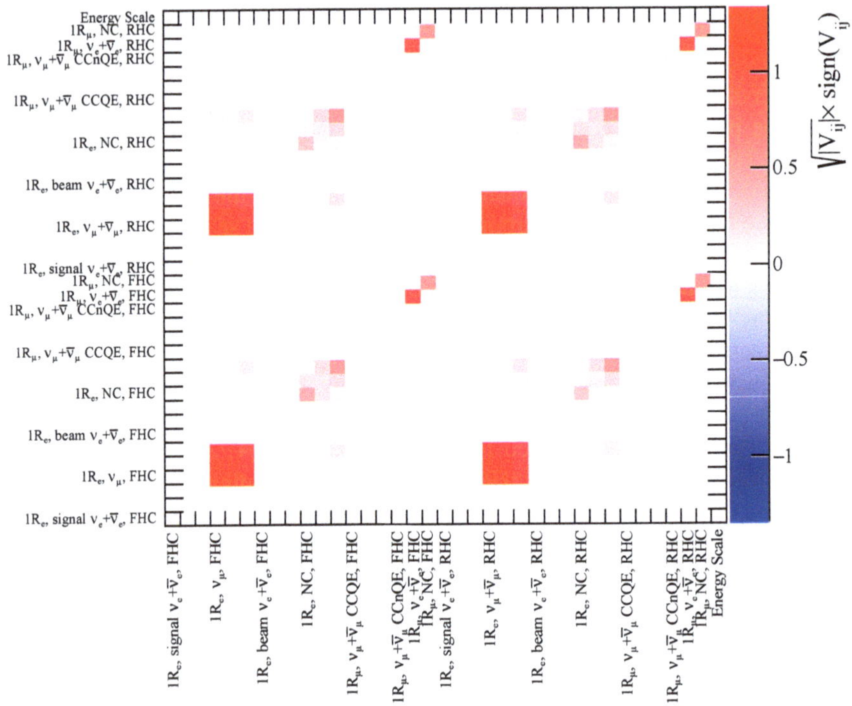

Fig. 5.4 The fractional covariance matrix used for the SK detector systematic parameters. The bin labels show where each group of bins (as given in the text) begins. Within each SK sub-sample, the bins iterate over reconstructed neutrino energy

5.4 Monte Carlo Predictions and Pre-fit Data/MC Comparison

5.4.1 Flux Tuning

The MC used for the prediction at both ND280 and SK was generated using the flux simulation described in Sect. 3.1.3, which was determined before the data was taken. However, small updates to the flux prediction from in-situ measurements during data-taking mean that the MC must be updated for this analysis. It is too computer-intensive to regenerate the entire MC every time there is an update to the prediction, so instead a tuning is used: weights are generated by the T2K beam group to reweight the MC to the desired prediction [26]. These are supplied in histograms binned in true neutrino energy, and are applied on an event-by-event basis in the analysis.

5.4.2 Tuning for Neutrino Interaction Model

As described in Sect. 3.4.1, the default T2K simulation is produced using the SF+MEC model for CCQE interactions, but the RFG+rel.RPA+MEC model has been found to fit both T2K and external CCQE data better [13]. Therefore, the MC must be reweighted to the RFG+rel.RPA+MEC model. It is possible to do this via reweighting, rather than needing to re-generate the MC, because the SF model predicts a broader lepton kinematic phase space than the RFG + rel.RPA model. The weights for this are generated and applied event-by-event to the MC.

Two additional tunings are needed. A normalisation is applied to double the number of NC 1γ events in the SK prediction, which seems to be a factor of 2 too low in NEUT [27], and a tuning is applied to events corresponding to CC Coherent interactions according to a comparison of the NEUT prediction to MINERνA data [17].

Once the flux and neutrino interaction model tunings have been applied to the MC it is termed the "pre-fit" MC prediction.

5.4.3 Tuning from Fits to ND280 Data

This analysis fits samples from ND280 and SK simultaneously, but other analyses on T2K take a different approach (as described in Appendix B). For these analyses, a fit is performed to only ND280 data to estimate the flux, cross-section, and ND280 detector systematic uncertainties, after which the flux and cross-section estimates and constraints are propagated to SK. These are taken as the input to an SK-only fit, which also includes SK detector systematic uncertainties and the neutrino oscillation parameters. The BANFF group on T2K is responsible for the ND280-only fit and produce a 'BANFF tuning' from the central values of the SK flux and cross-section parameters after the ND280 fit. This is largely irrelevant for this analysis, but the BANFF tuning is sometimes used to generate 'pre-SK-fit' predictions which are comparable with the other analyses, in order to provide a cross-check of the framework.

5.4.4 ND280 MC Prediction

Table 5.4 gives the number of events in the ND280 data and pre-fit MC samples, taken from [8]. The predicted pre-fit momentum distributions, integrated over $\cos\theta$ and with the data overlaid, are shown in Figs. 5.5 and 5.6 for FGD1 and Figs. 5.7 and 5.8 for FGD2. The prediction is shown as a stacked histogram by interaction mode: charged-current quasi-elastic scattering (CCQE), charged-current 2p-2h scattering (CC 2p-2h), charged-current resonant single pion production (CC Res 1π),

Table 5.4 Number of data events in each of the ND280 samples, compared to the pre-fit MC prediction

	CC0π	CC1π	CC Other	$\bar{\nu}_\mu$ CC 1	$\bar{\nu}_\mu$ CC N	ν_μ CC 1	ν_μ CC N
FGD1							
Data	17354	3984	4220	2663	775	989	1001
Pre-fit MC	16950.81	4460.15	4009.78	2707.65	797.73	938.13	995.33
Data/MC Ratio	1.02	0.89	1.05	0.98	0.97	1.05	1.01
FGD2							
Data	17650	3383	4118	2762	737	980	936
Pre-fit MC	17211.71	3616.62	3626.56	2729.88	804.45	943.90	916.61
Data/MC Ratio	1.03	0.94	1.14	1.01	0.92	1.04	1.02

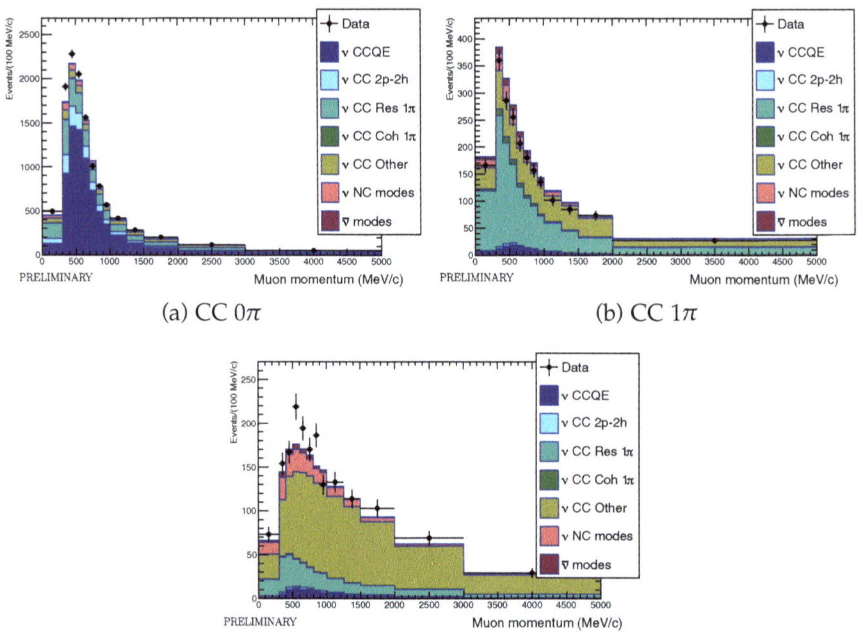

(a) CC 0π

(b) CC 1π

(c) CC Other

Fig. 5.5 ND280 pre-fit MC and data for the FHC samples in FGD1. Copyright by the T2K Collaboration

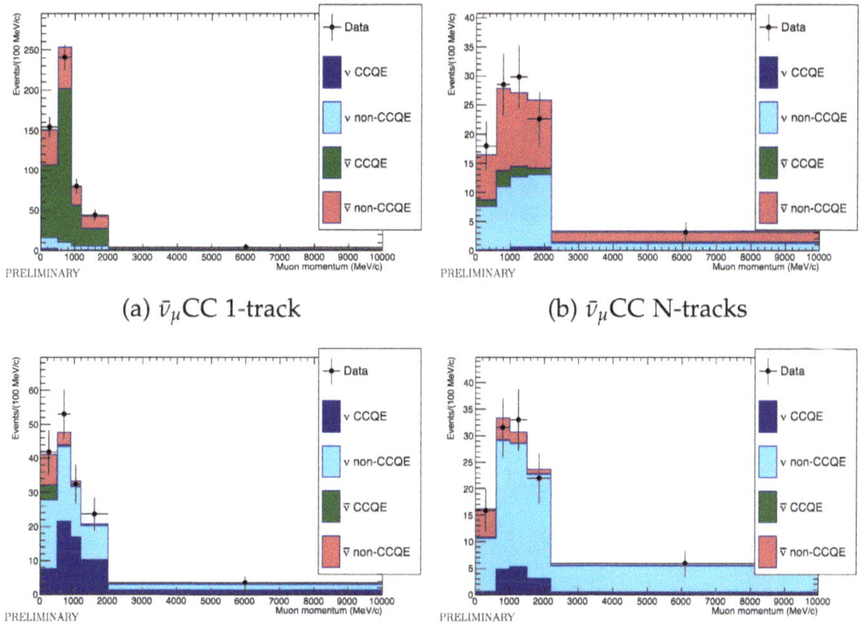

(a) $\bar{\nu}_\mu$CC 1-track

(b) $\bar{\nu}_\mu$CC N-tracks

(c) ν_μCC 1-track

(d) ν_μCC N-tracks

Fig. 5.6 ND280 pre-fit MC and data for the RHC samples in FGD1. Copyright by the T2K Collaboration

charged-current coherent single pion production (CC Coh 1π), other charged-current interactions (CC Other), and neutral current scattering (NC modes).

The ν_μ CC 0π and ν_μ CC Other neutrino-mode samples contain more data events than predicted and the ν_μ CC 1π neutrino-mode sample contains fewer data events than predicted. The consequence of this in the fit is that the flux normalisation parameters are generally increased by 10–15% (in order to increase the number of predicted events in the ν_μ CC 0π and ν_μ CC Other samples), and the cross-section parameters that correspond to resonance interactions are moved to smaller values (in order to reduce the number of predicted events in the ν_μ CC 1π sample). The results of the data fit in terms of the systematic parameters are shown in Sect. 6.1.

5.4.5 *Super-Kamiokande MC Prediction*

The SK MC is provided by the T2K-SK group. For each selection ($1R_\mu$ or $1R_e$ in neutrino or antineutrino mode), the MC is generated for six true neutrino flavours: beam ν_μ, beam $\bar{\nu}_\mu$, beam ν_e, beam $\bar{\nu}_e$, signal ν_e, and signal $\bar{\nu}_e$. The four beam samples are each generated according to the expected flux of that type of particle in the beam.

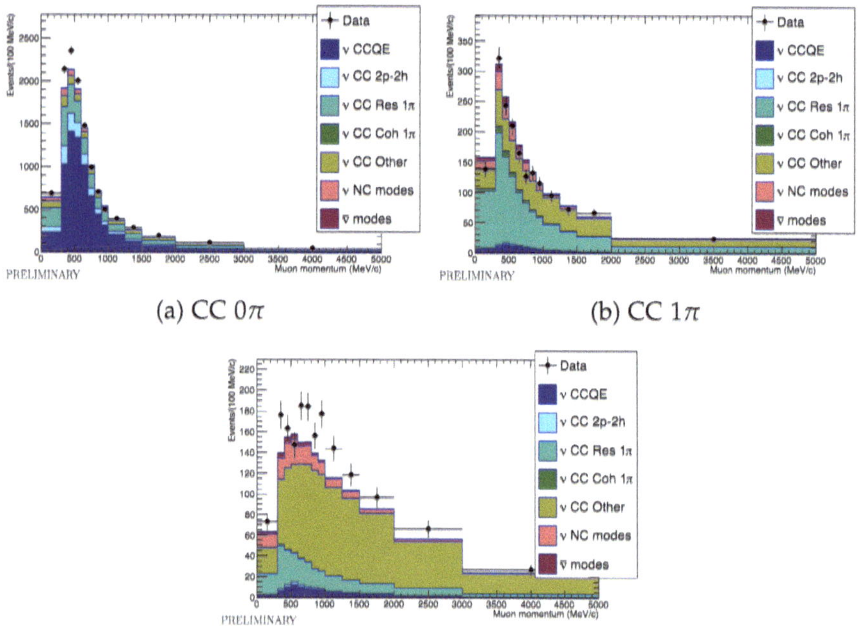

(a) CC 0π

(b) CC 1π

(c) CC Other

Fig. 5.7 ND280 pre-fit MC and data for the FHC samples in FGD2. Copyright by the T2K Collaboration

The final two categories, signal ν_e and signal $\bar{\nu}_e$, contain ν_e and $\bar{\nu}_e$ events respectively, but are generated according to the ν_μ or $\bar{\nu}_\mu$ flux predictions – this corresponds to a 100% $\nu_\mu \rightarrow \nu_e$ (or $\bar{\nu}_\mu \rightarrow \bar{\nu}_e$) oscillation. Events are then weighted according to the desired oscillation probability for ν_e or $\bar{\nu}_e$ appearance. Events in the four 'beam' samples are also weighted according to the oscillation survival probability.

The predicted numbers of events in the SK data set are presented both for the oscillated and unoscillated case. The 'oscillated' predictions have been found using the following oscillation parameters, with $\sin^2\theta_{23}$, Δm^2_{32}, and δ_{CP} taken from the result of the T2K ν_e+ν_μ joint fit including the reactor constraint [10], and $\sin^2\theta_{12}$, $\sin^2\theta_{13}$, and Δm^2_{21} taken from the 2015 PDG [2]:

- $\sin^2\theta_{12} = 0.304$
- $\sin^2\theta_{23} = 0.528$
- $\sin^2\theta_{13} = 0.0217$
- $\Delta m^2_{21} = 7.53\times10^{-5}\text{eV}^2$
- $\Delta m^2_{32} = 2.509\times10^{-3}\text{eV}^2$
- $\delta_{CP} = -1.601$

where the positive sign of Δm^2_{32} indicates that the normal mass hierarchy is assumed. Table 5.5 shows the predicted numbers of events in each sample for the pre-fit MC, broken down by true neutrino flavour (where 'true' neutrino flavour represents the

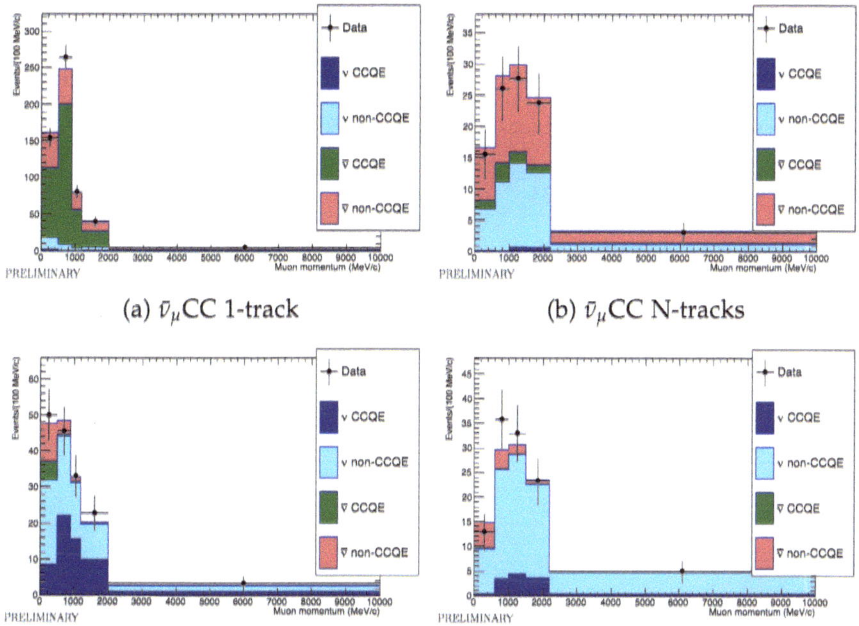

(a) $\bar{\nu}_\mu$CC 1-track (b) $\bar{\nu}_\mu$CC N-tracks

(c) ν_μCC 1-track (d) ν_μCC N-tracks

Fig. 5.8 ND280 pre-fit MC and data for the RHC samples in FGD2. Copyright by the T2K Collaboration

flavour of the neutrino in the MC, as opposed to the selection in which the interaction falls). No selection is perfect, and in each of the four SK samples we expect some contamination from other neutrino flavours. In particular, contamination from instrinsic ν_e or $\bar{\nu}_e$ in the beam will form a significant background in the $1R_e$ selections. To account for this, contributions from all six neutrino categories described above are considered for each selection. We also consider neutral-current interactions in all six categories (the oscillated prediction uses the unoscillated predicted neutral-current interaction rate, in order to correctly account for ν_τ interactions). Table 5.6 summarises the total number of data events and predicted pre-fit MC events, and their ratio, for each of the four samples.

Figure 5.9 shows the oscillated and unoscillated pre-fit MC prediction, and the data, as a function of reconstructed neutrino or antineutrino energy. Figure 5.10 shows the predicted reconstructed energy spectrum at SK as a stacked histogram by interaction mode: charged-current quasi-elastic scattering (CCQE), charged-current 2p-2h scattering (2p-2h), charged-current resonant single pion production (CC1π), charged-current resonant multiple pion production (CC nπ), charged-current coherent pion production (CC Coherent), other charged-current interactions (CC Other), neutral current π^0 production (NC π^0), neutral current single pion resonant pro-

Table 5.5 Summary of the number of events in each sample at SK from the pre-fit MC prediction, broken down by true neutrino flavour. The prediction is given for the total number of events due to each type of neutrino in each selection, including neutral-current interactions, when ignoring neutrino oscillation ('unosc.') and when considering neutrino oscillation ('osc.'). The oscillation parameters used for the 'osc.' predictions are given in the text

SK sample		ν_μ	ν_e	$\bar{\nu}_\mu$	$\bar{\nu}_e$	ν_e signal	$\bar{\nu}_e$ signal	Total
FHC $1R_\mu$	Unosc.	437.88	0.33	13.74	0.03	0.00	0.00	451.97
	Osc.	113.44	0.33	7.78	0.03	0.32	0.00	121.90
FHC $1R_e$	Unosc.	2.11	3.18	0.08	0.15	0.00	0.00	5.52
	Osc.	1.65	2.96	0.08	0.15	19.59	0.14	24.56
RHC $1R_\mu$	Unosc.	47.13	0.11	130.21	0.07	0.00	0.00	177.52
	Osc.	25.97	0.11	36.28	0.07	0.02	0.04	62.48
RHC $1R_e$	Unosc.	0.32	0.57	0.51	0.92	0.00	0.00	2.32
	Osc.	0.30	0.53	0.47	0.87	0.95	2.70	5.82

Table 5.6 Number of data events in each of the SK samples compared to the oscillated pre-fit MC prediction

	FHC $1R_\mu$	FHC $1R_e$	RHC $1R_\mu$	RHC $1R_e$
Data	125	32	66	4
Pre-fit MC	121.90	24.56	62.48	5.82
Data/MC Ratio	1.03	1.30	1.06	0.69

duction (NC $\pi^{\pm}$), neutral current coherent pion production (NC Coherent), neutral current γ production (NC 1γ), and other neutral current interactions (NC Other).

5.4.6 *Effect of Systematic Uncertainties on SK Event-Rate Prediction*

Tables 5.7 and 5.8 show the impact of the systematic parameters on the SK samples included in this analysis, as an uncertainty on the total number of events in each sample. Table 5.9 shows the systematic uncertainties on the ν-mode/$\bar{\nu}$-mode ratio for the number of events in the $1R_\mu$ and $1R_e$ samples, which will impact the analysis' ability to measure δ_{CP}. The 'pre-fit' uncertainty is calculated by throwing each set of parameters from Gaussian distributions according to the prior covariance matrices introduced in previous sections (including correlations) 10,000 times. The quantity $\Delta N_{SK}/N_{SK}$ is calculated for each sample, where N_{SK} is the mean and ΔN_{SK} is the RMS of the distribution of the number of predicted events.

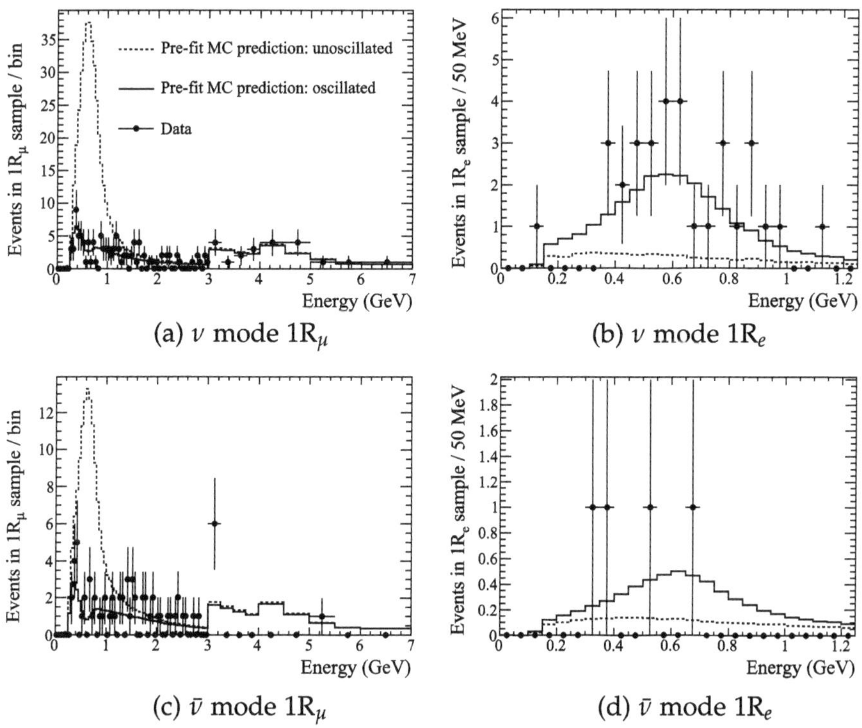

(a) ν mode $1R_\mu$ (b) ν mode $1R_e$

(c) $\bar{\nu}$ mode $1R_\mu$ (d) $\bar{\nu}$ mode $1R_e$

Fig. 5.9 Oscillated (*solid*) and unoscillated (*dashed*) reconstructed energy spectra for all four Super-K samples, using the pre-fit SK MC. The $1R_\mu$ plots are zoomed in on the range 0–7 GeV, as no data are found outside that range, although the fit window extends up to 30 GeV. The data are shown in *black*, and the oscillation parameters used are given in the text. Note that the y-axis scale for the $1R_\mu$ plots shows 'Events per bin', and the binning changes from 50 MeV-width bins to wider, non-uniform, bins at 3 GeV

It is, of course, impossible to define Gaussian pre-fit distributions for those cross-section parameters for which we use flat priors. Instead they are thrown from flat distributions in the following ranges, which were deemed to be reasonable by the NIWG group:

- M_A^{QE}: 0.8–1.2 (multiplies 1.21 GeV)
- 2p–2h (C), 2p–2h (O): 0.0–2.5
- 2p–2h $\bar{\nu}/\nu$: 0.4–1.6
- p_F (C): 0.92–1.27
- p_F (O): 0.89–1.22

A 'post-ND fit' uncertainty in the number of events at SK is also shown in the tables. This is calculated using a posterior predictive method from the posterior of a fit to ND280 data only. 10,000 random points from the fit posterior are chosen, and the SK MC tuned to the values of the flux and cross-section parameters of that point. Because the SK detector uncertainties do not factor in an ND280-only fit, these are

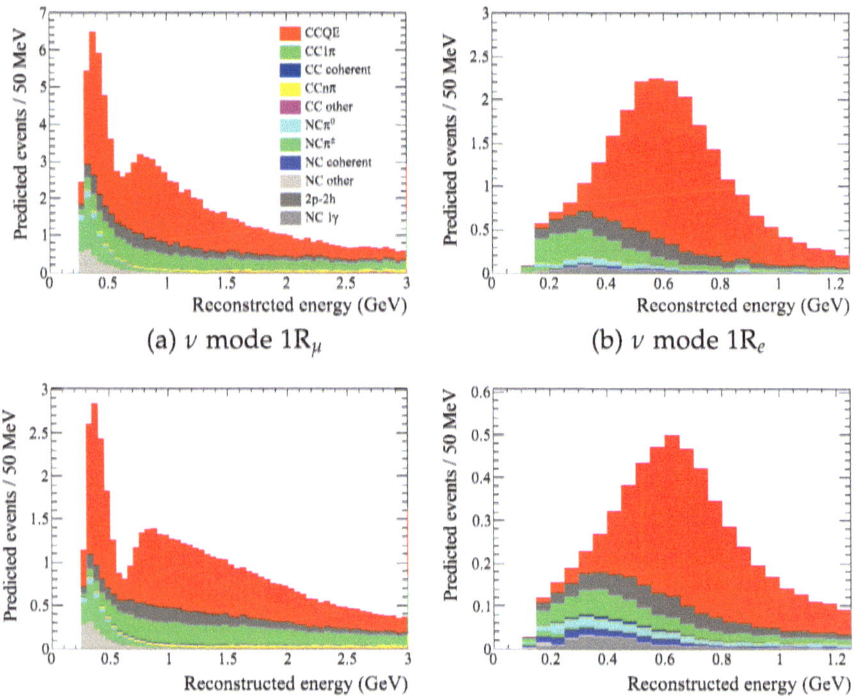

(a) ν mode $1R_\mu$

(b) ν mode $1R_e$

(c) $\bar{\nu}$ mode $1R_\mu$

(d) $\bar{\nu}$ mode $1R_e$

Fig. 5.10 Reconstructed energy spectra for all four SK samples as a stacked histogram by interaction mode, using the pre-fit SK MC. The $1R_\mu$ plots are zoomed in around the oscillation dip for clarity, and the oscillation parameters used are given in the text

thrown from the covariance matrix, as for the pre-fit uncertainty. $\Delta N_{SK}/N_{SK}$ is then calculated from the distribution of the predicted number of events at each of these 10,000 points.

The tables show that fitting the ND280 data alone reduces the total uncertainty due to flux and cross-section uncertainties to below 3% for the ν-mode $1R_e$, ν-mode $1R_\mu$, and $\bar{\nu}$-mode $1R_\mu$ samples, and around 4.5% for the $\bar{\nu}$-mode $1R_e$ sample. This is often smaller than the uncertainty due to either flux or cross-section systematic parameters alone because of anti-correlations between parameters. Overall, the total uncertainty in the predicted number of events is reduced from around 12–13% to around 5% for each of the four SK samples as a result of the ND280 data.

Table 5.7 Fractional uncertainty on the number of events predicted in the SK ν-mode samples due to different sources of systematic uncertainty, before and after a fit to the near detector (ND) data

Systematic	$1R_\mu$		$1R_e$	
	$\Delta N_{SK}/N_{SK}$ Before ND fit (%)	$\Delta N_{SK}/N_{SK}$ After ND fit (%)	$\Delta N_{SK}/N_{SK}$ Before ND fit (%)	$\Delta N_{SK}/N_{SK}$ After ND fit (%)
Flux	7.7	3.1	8.8	3.2
Cross-section	7.6	3.8	7.1	4.7
Flux and cross-section	10.9	2.5	11.4	2.7
FSI/SI at SK	1.8		2.5	
SK detector	4.6		2.5	
Total	12.1	4.9	11.9	5.2

Table 5.8 Fractional uncertainty on the number of events predicted in the SK $\bar{\nu}$-mode samples due to different sources of systematic uncertainty, before and after a fit to the near detector (ND) data

Systematic	$1R_\mu$		$1R_e$	
	$\Delta N_{SK}/N_{SK}$ Before ND fit (%)	$\Delta N_{SK}/N_{SK}$ After ND fit (%)	$\Delta N_{SK}/N_{SK}$ Before ND fit (%)	$\Delta N_{SK}/N_{SK}$ After ND fit (%)
Flux	7.1	3.3	7.9	3.4
Cross-section	9.3	4.0	10.1	5.4
Flux and cross-section	11.7	2.9	12.9	4.5
FSI/SI at SK	2.4		2.4	
SK detector	3.9		3.1	
Total	12.6	4.8	13.4	5.7

Table 5.9 Fractional uncertainty on the ratio of the number of events predicted in the SK ν-mode and $\bar{\nu}$-mode samples due to different sources of systematic uncertainty, before and after a fit to the near detector (ND) data

Systematic	$1R_\mu$		$1R_e$	
	$\Delta N_{SK}/N_{SK}$ Before ND fit (%)	$\Delta N_{SK}/N_{SK}$ After ND fit (%)	$\Delta N_{SK}/N_{SK}$ Before ND fit (%)	$\Delta N_{SK}/N_{SK}$ After ND fit (%)
Flux	4.4	1.7	3.3	1.7
Cross-section	4.2	1.7	7.0	4.2
Flux and cross-section	6.0	1.7	7.7	4.1
FSI/SI at SK	3.0		3.3	
SK detector	0.8		1.5	
Total	6.7	3.1	8.4	5.4

5.5 Fitter Validation and Sensitivity

Many tests were conducted to validate the framework for this analysis, including cross-checks against the BANFF group and other oscillation analysis groups on T2K, and space limitations mean that it is not possible to describe all validations in detail here. Instead, a selection of representative validation tests will be discussed, and for more details the reader is directed to [1, 8].

The code to handle the ND280 data, prediction, and systematic parameters was updated for this analysis by Dr. A. Kaboth. Multiple validations of this implentation, including comparisons of ND280-only fits against the BANFF group, were conducted by him and are described in [8]. To validate the SK side of the framework, the nominal number of predicted events in each SK sample, presented in Sect. 5.4.5, and the uncertainty tables presented in Sect. 5.4.6 were compared between the three T2K oscillation groups (called MaCh3—described in this thesis—VaLOR [28], and P-Theta [29]). The three oscillation groups use the same inputs but have developed fitting frameworks independently, so comparisons between the groups are useful to validate the frameworks and check for bugs. The main similarities and differences between the three analyses are described in Appendix B. The effect on each of the SK samples of a $\pm 1\sigma$ or 3σ variation in each of the systematic parameters was also checked, and good agreement was found in all three cases.

These tests show that the implementation of the MC and systematic parameters used for this analysis is working as we expect. Now it remains to validate the oscillation fit machinery. Two methods are shown for this: a fit to an Asimov data set, described in Sect. 5.5.1, and a scan of the Bayesian log-likelihood (LLH) as a function of the oscillation parameters, described in Sect. 5.5.2. Both validations use the oscillation parameters defined in Sect. 5.4.5 as their input.

5.5.1 Asimov Fits

An Asimov fake data set is defined to be a toy experiment generated from the MC prediction in such a way that there are no statistical fluctuations. This is achieved by reweighting the MC to chosen values of the oscillation and systematic parameters, along with the required protons on target. Then, instead of drawing randomly from the prediction, the prediction is considered as the data set—that is, the "data" *is* the MC prediction. This is obviously unphysical, as it can result in non-integer numbers of "data" events, but it gives a data set free from statistical fluctuations which—when fit—should result in parameters free from bias.

An Asimov fake data fit may be more useful for fitter validation than a fake data set with realistic statistical variations, because if the fitter is working correctly we know exactly what result we expect to see. If an anomalous fit result is found, it must be due to a problem in the fitter, not due to statistical fluctuations. It also has the advantage that it can be used to estimate the sensitivity of an analysis: the contours from an Asimov fit will show us the maximum precision with which we expect to be able to measure the oscillation parameters.

To validate the joint fit machinery, two fits were completed, both using the oscillation parameters defined in Sect. 5.4.5: one with the reactor constraint on $\sin^2 2\theta_{13}$, and one without. These are not actually true Asimov fake data fits: instead, they are fits to ND280 data and SK Asimov fake data, where the SK Asimov fake data has the BANFF tuning applied (as defined in Sect. 5.4.3). This enables us to compare results with the other oscillation groups, as this ND280 data/SK Asimov fit is equivalent to an SK-only Asimov fit using the BANFF ND280 data fit result as an input.

5.5.1.1 Asimov Fit: T2K only

Figure 5.11 shows the posterior probability distribution and credible intervals from an Asimov fit without the reactor constraint. The 2D histogram shows the number of steps in the MCMC output at each point in 2D parameter space, which is proportional to the posterior probability density (the units of the z-axis are arbitrary, but darker shades represent larger posterior density). The black lines show the 90% (solid line) and 68% (dashed line) credible intervals, the white triangle marks the mode of the 2D marginal posterior, and the grey square marks the mode of the 4D marginal posterior. The left-hand plot shows the marginal posterior and 2D credible intervals in $\sin^2\theta_{23}$–Δm^2_{32} and the right-hand figure shows the same for $\sin^2\theta_{13}$–δ_{CP}.

Because the chain spans both mass hierarchies in a single fit, the 68% and 90% credible intervals include steps from both mass hierarchies. The 90% credible interval contains 90% of the total posterior probability density, which includes steps with $\Delta m^2_{32}>0$ and $\Delta m^2_{32}<0$. This takes into account the relative probabilities of the two

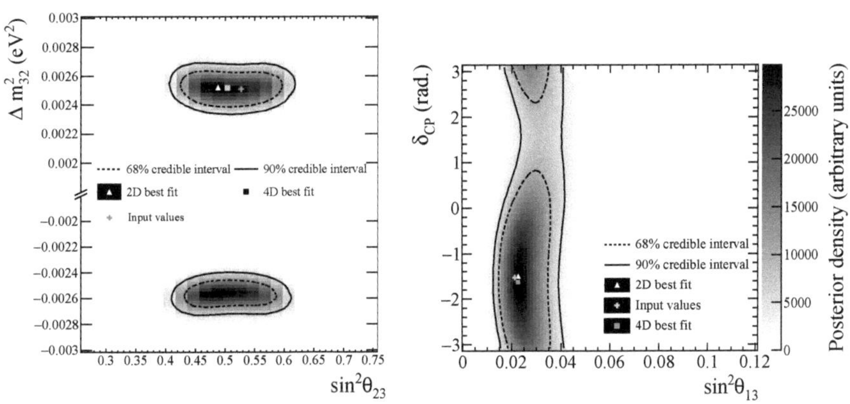

Fig. 5.11 Posterior distribution and credible intervals from an Asimov fit without the reactor constraint. The input values, 2D posterior mode (labelled "2D best fit"), and 4D posterior mode (labelled "4D best fit") are also shown. *Left* $\sin^2\theta_{23}$–Δm^2_{32}. *Right* $\sin^2\theta_{13}$–δ_{CP}

mass hierarchies in the posterior, and will produce a different contour than the 90% credible interval considering normal hierarchy steps only (both in $\sin^2\theta_{23}$–Δm^2_{32} and $\sin^2\theta_{13}$–δ_{CP} space). The contours in $\sin^2\theta_{23}$–Δm^2_{32} show a discontinuity across the choice of mass hierarchy, and therefore the relative size of the contours in normal and inverted hierarchy can be used to estimate the relative probabilities. In this fit the posterior probability for normal and inverted hierarchy is roughly equal, and so the contours in $\sin^2\theta_{23}$–Δm^2_{32} are a similar size in both hierarchies. If the normal hierarchy—for example—were strongly preferred, the contour would be larger for $\Delta m^2_{32}>0$ than for $\Delta m^2_{32}<0$.

In $\sin^2\theta_{13}$–δ_{CP}, the mode of both the 2D and 4D marginal posterior is very close to the input values. The fact that the 2D and 4D posterior modes are so close together implies that there is not a strong marginalisation effect due to either of the oscillation parameters $\sin^2\theta_{23}$ and Δm^2_{32} (recall that the marginalisation of these parameters is the only difference between the 2D and 4D marginal posterior). The fact that both of these are close to the input value further implies that there is no strong marginalisation effect due the other nuisance parameters. The much more interesting aspect of this figure, however, is that it shows sensitivity to δ_{CP} in a T2K-only fit for the first time. The previous T2K ν_e+ν_μ fit [10] (the results of which are reproduced for comparison to the ν+$\bar{\nu}$ data fit in Sect. 6.2) had no power to constrain δ_{CP} when fitting without the reactor constraint. Adding the antineutrino data gives sensitivity to δ_{CP} when fitting T2K data alone—we expect a closed 68% contour which excludes a region around $\delta_{CP}\simeq\pi/2$. This highlights the exciting prospect that this analysis, and future statistical updates, may be capable of constraining δ_{CP} without relying on external data constraints of $\sin^2\theta_{13}$.

In $\sin^2\theta_{23}$–Δm^2_{32}, we see a larger difference between the 2D input value and the two best-fit points. The mode of the 2D posterior is at a somewhat lower value of $\sin^2\theta_{23}$ than the input, and the mode of the 4D marginal posterior is closer to the input value, suggesting that the shift in the 2D mode comes from marginalisation over either $\sin^2\theta_{13}$ or δ_{CP}. It seems that this is due to a marginalisation effect when marginalising over $\sin^2\theta_{13}$, as shown in Fig. 5.12. We see the posterior distribution in $\sin^2\theta_{13}$ is wider at lower values of $\sin^2\theta_{23}$, which may explain the push to low

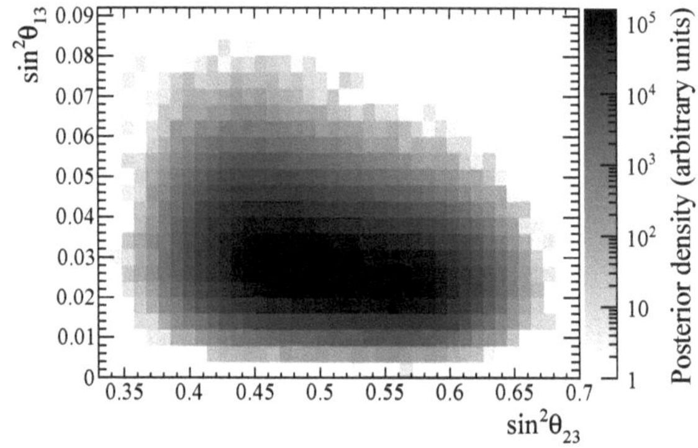

Fig. 5.12 2D posterior in $\sin^2\theta_{23}$–$\sin^2\theta_{13}$ from an Asimov fit without the reactor constraint (note the log scale in the z axis). The shape of this distribution leads to a marginalisation effect in the 2D $\sin^2\theta_{23}$–Δm^2_{32} best-fit point

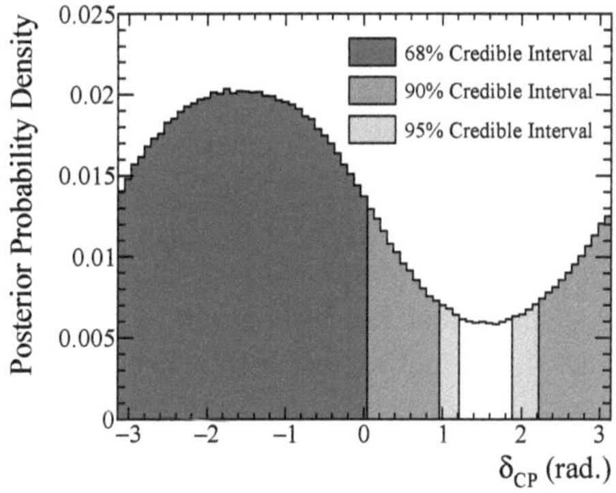

Fig. 5.13 1D posterior in δ_{CP} from an Asimov fit without the reactor constraint. The 1D 68, 90, and 95% credible intervals are also shown

$\sin^2\theta_{23}$ when $\sin^2\theta_{13}$ is marginalised to produce the 2D posterior. The fact that the 4D mode is still not identical to the input values indicates additional marginalisation effects from the other parameters.

The 1D posterior and 68, 90, and 95% 1D credible intervals as a function of δ_{CP} is shown in Fig. 5.13. The distribution peaks around the input value, as we expect, and excludes the entire region $\delta_{CP} = 0$ to π from the 68% credible interval (again, a first for a T2K-only sensitivity).

The Markov chain also provides an interesting and natural way to compare the posterior probabilities of the two mass hierarchies and the two octants of θ_{23}. The posterior probability (given the data and prior assumptions) that a given combination of mass hierarchy and octant is correct is simply the proportion of steps of the chain in that region of phase space. Table 5.10 provides these probabilities for the two choices of mass hierarchy and two choices of octant. With an input value of the oscillation parameters in the normal hierarchy and (just) in the upper octant, we see 57% of the posterior probability in normal hierarchy and 55% of the posterior probability in the upper octant.

Because we chose equal priors on both the mass hierarchies (P(NH)=P(IH)=0.5), we can calculate a Bayes factor for the mass hierarchy from the numbers in Table 5.10. The Bayes factor is the ratio of marginal likelihoods of two hypotheses, and in the special case of equal priors on the two hypotheses it is also equal to the ratio of posterior probabilities. From Table 5.10, the posterior probability is 57% for normal hierarchy (NH) and 43% for inverted hierarchy (IH), giving a Bayes factor B(NH/IH)

Table 5.10 Model comparison probabilities for normal and inverted hierarchies, as well as upper and lower octants, from the posterior of the T2K-only Asimov fit

	$\sin^2\theta_{23} < 0.5$	$\sin^2\theta_{23} > 0.5$	Sum
IH ($\Delta m^2_{32} < 0$)	0.20	0.23	0.43
NH ($\Delta m^2_{32} > 0$)	0.25	0.32	0.57
Sum	0.45	0.55	1

= 1.33. In a similar way, we find the Bayes factor for the upper octant, $B(\sin^2\theta_{23} > 0.5/\sin^2\theta_{23} < 0.5)$ to be 1.22. Therefore, we should expect to have little sensitivity to either the octant or the mass hierarchy from the data, assuming the oscillation parameters used for the input here.

As further validation, the results of this fit were compared with the two other T2K oscillation groups. Because these groups use a hybrid frequentist-Bayesian method which marginalises systematics but produces confidence intervals, constant-$\Delta\chi^2$ confidence level contours were made (following the procedure described in Sect. 4.4.3) for the comparison. In addition, the other analyses do not marginalise over the mass hierarchy, so steps in only one mass hierarchy (either normal or inverted) were selected to make the contours. Note that these 2D confidence intervals are calculated with respect to the 2D best-fit point (i.e. the mode of the 2D posterior probability distribution), so the 4D best-fit point is not shown.

As an example, Fig. 5.14 shows the comparison of constant-$\Delta\chi^2$ confidence intervals with the other groups, and the comparison of $\Delta\chi^2$ as a function of δ_{CP}, both for normal mass hierarchy only (the VaLOR contours in 2D $\sin^2\theta_{23}$–Δm^2_{32} and the P-Theta 1D δ_{CP} distributions were not available, so those comparisons are not shown). These figures show very good agreement between the analyses: there are some differences in the 2D best-fit points, but the fact that the confidence intervals agree so well implies that the χ^2 for all three best-fit points is very similar. Where there are differences in the best-fit points, they arise because the distribution around the point is very flat (Fig. 5.11 shows that the posterior distribution is very flat close to the best-fit points in both $\sin^2\theta_{23}$–Δm^2_{32} and in $\sin^2\theta_{13}$–δ_{CP}), and the difference that we see in best-fit points does not produce noticeably different constant-$\Delta\chi^2$ confidence intervals. There are also some small differences in the 1D $\Delta\chi^2$ distributions, but these are on the order of $\Delta\chi^2 \simeq 0.1$. The inverted-hierarchy confidence intervals and 1D $\Delta\chi^2$ distribution are not shown for brevity, but display equally good agreement between the three groups.

5.5.1.2 Asimov Fit: T2K + Reactor

Figure 5.15 shows the posterior probability distribution and credible intervals from an Asimov fit with the reactor constraint on $\sin^2 2\theta_{13}$. In $\sin^2\theta_{13}$–δ_{CP}, the mode of the 2D marginal posterior and 4D marginal posterior is at the input value. We see a closed 68% and 90% contour in δ_{CP}, which means that we may have sensitivity to exclude some values of δ_{CP} at 90% probability.

In $\sin^2\theta_{23}$–Δm^2_{32}, the mode of the 2D posterior is slightly higher than the input value. As in the fit without the reactor constraint, the mode of the 4D marginal posterior is much closer to the input values, suggesting that the difference is due to marginalisation over either $\sin^2\theta_{13}$ or δ_{CP}. Because $\sin^2\theta_{13}$ is so tightly constrained by the reactor constraint, it is reasonable to expect that this must be a marginalisation effect over δ_{CP}. This is indeed the case, as shown in Fig. 5.16: the 2D posterior probability distribution as a function of $\sin^2\theta_{23}$–δ_{CP} has a larger spread in δ_{CP} at high

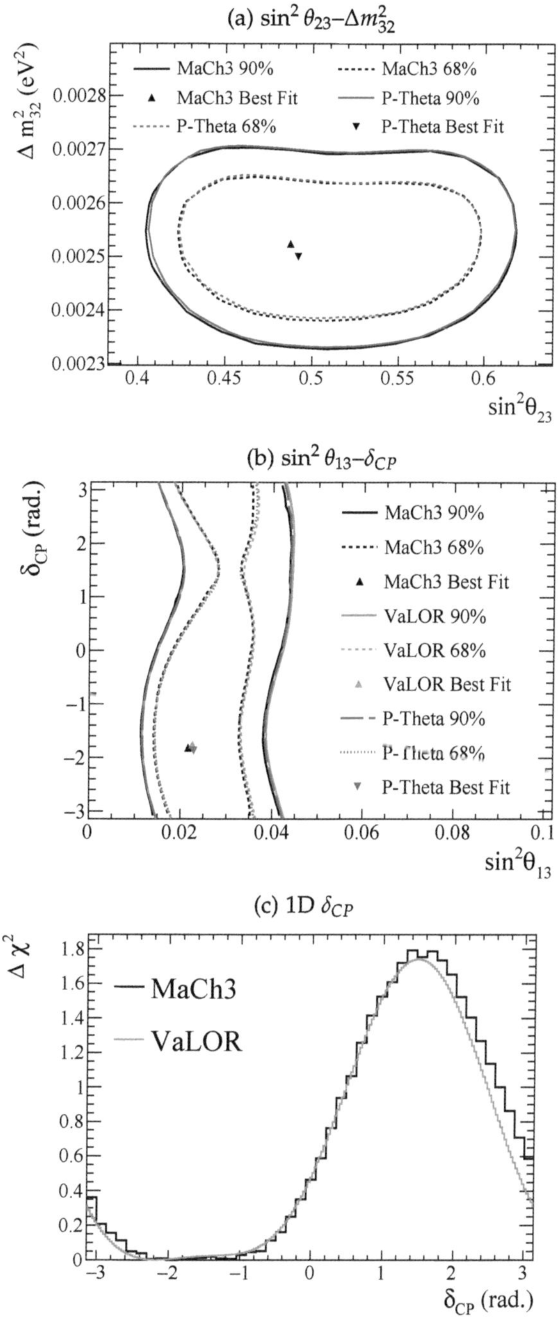

Fig. 5.14 Comparison of normal-hierarchy confidence intervals and 1D $\Delta\chi^2$ as a function of δ_{CP} between this analysis and other T2K analyses, from an Asimov fit without the reactor constraint

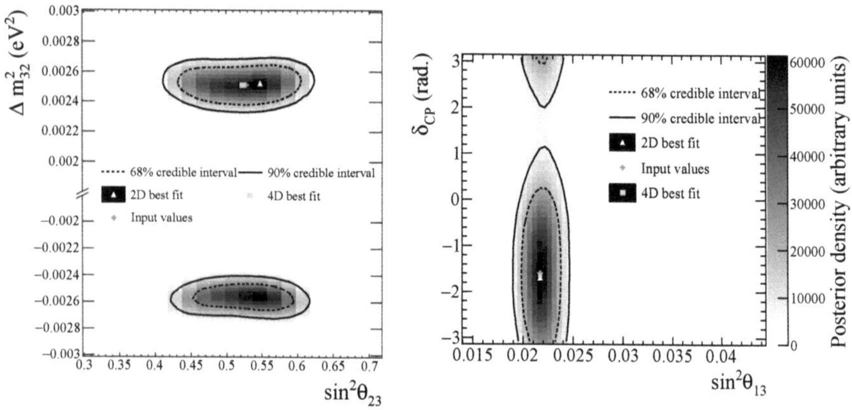

Fig. 5.15 Posterior distribution and credible intervals from an Asimov fit with the reactor constraint on $\sin^2 2\theta_{13}$. The input values, 2D posterior mode (labelled “2D best fit”), and 4D posterior mode (labelled “4D best fit”) are also shown. *Left* $\sin^2\theta_{23}$–Δm^2_{32}. *Right* $\sin^2\theta_{13}$–δ_{CP}

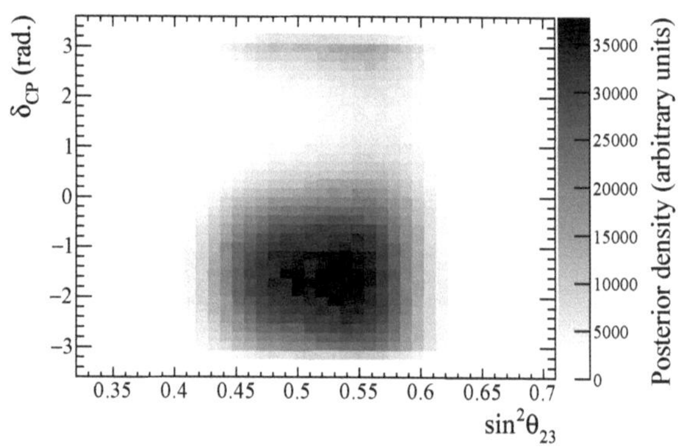

Fig. 5.16 2D posterior in $\sin^2\theta_{23}$–δ_{CP} from an Asimov fit with the reactor constraint on $\sin^2 2\theta_{13}$. The shape of this distribution leads to a marginalisation effect in the 2D $\sin^2\theta_{23}$–Δm^2_{32} best-fit point

values of $\sin^2\theta_{23}$ than at low values. Therefore, when the distribution is marginalised over δ_{CP}, it will shift towards high $\sin^2\theta_{23}$.

Figure 5.17 shows the 1D posterior and credible intervals as a function of δ_{CP}. The distribution is much more sharply peaked than for the fit without the reactor constraint, giving smaller credible intervals and better sensitivity to δ_{CP}. For this fit—with the reactor constraint—it seems that we may have sensitivity to exclude δ_{CP}=0 and $\delta_{CP}=\pm\pi$ at 68%, although not at 90% or 95% probability.

Table 5.11 shows the posterior probability for each combination of mass hierarchy and octant. For this fit, we see 63% of the posterior probability in normal hierarchy and 66% of the posterior probability in the upper octant. This corresponds to a Bayes factor B(NH/IH) of 1.70, and $B(\sin^2\theta_{23} > 0.5/\sin^2\theta_{23} < 0.5)$=1.94. As in the previous section, this implies that we will not have strong sensitivity to either the octant or the mass hierarchy from the data, assuming the oscillation parameters used for the input here.

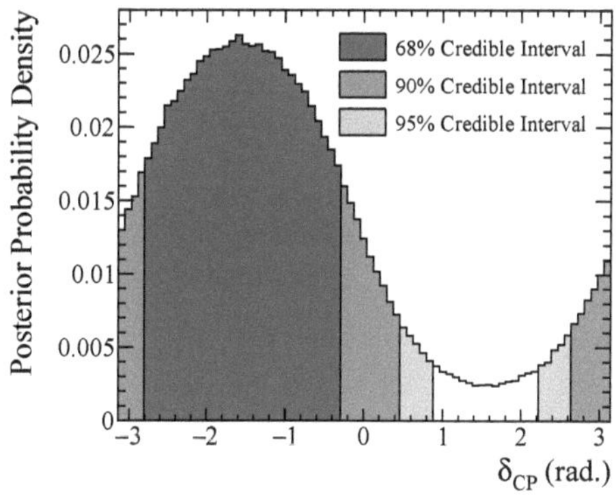

Fig. 5.17 1D posterior in δ_{CP} from an Asimov fit with the reactor constraint on $\sin^2 2\theta_{13}$. The 1D 68, 90, and 95% credible intervals are also shown

Table 5.11 Model comparison probabilities for normal and inverted hierarchies, as well as upper and lower octants, from the posterior of the T2K+Reactor Asimov fit

	$\sin^2\theta_{23} < 0.5$	$\sin^2\theta_{23} > 0.5$	Sum
IH ($\Delta m^2_{32} < 0$)	0.12	0.25	0.37
NH ($\Delta m^2_{32} > 0$)	0.22	0.41	0.63
Sum	0.34	0.66	1

Figure 5.18 shows the comparison of constant-$\Delta\chi^2$ confidence level contours and $\Delta\chi^2$ as a function of δ_{CP} with the VaLOR and P-Theta groups, again for normal hierarchy only (the P-Theta results were not available for $\Delta\chi^2$ as a function of δ_{CP}, so are not shown). As in the previous section, we see excellent agreement in the contours between all three groups. The inverted-hierarchy intervals and 1D $\Delta\chi^2$ distribution are not shown here, but display equally close agreement.

5.5.2 *Log-Likelihood Scan*

As we saw in Sect. 5.5.1, marginalisation effects mean that it is not always possible to determine from Asimov fits (or, indeed, other fake data fits) whether a fitter is biased. A non-biased fitter may still find a best-fit point which differs from the input values simply due to correlations with marginalised parameters. Therefore it is also important to test that the likelihood evaluated by the fitter is not biased.

The most straightforward way to conduct this test is to investigate the likelihood at different values of the interesting oscillation parameters $\sin^2\theta_{23}$, $\sin^2\theta_{13}$, Δm^2_{23}, and δ_{CP}, with all other parameters fixed. When the nuisance parameters are fixed, the maximum likelihood should be at the input value as there can be no marginalisation effects. If it is not, there is a bias in the fit machinery.

An Asimov fake data set was generated with the oscillation parameters defined in Sect. 5.4.5 and nominal systematic parameters, and a grid was defined in either

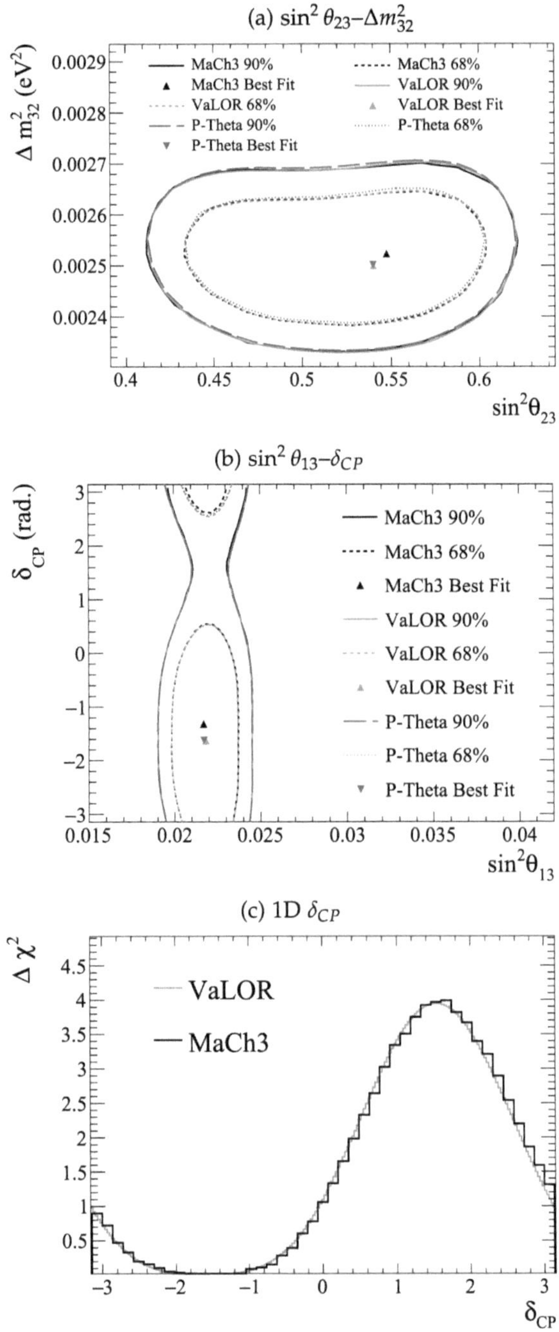

Fig. 5.18 Comparison of normal-hierarchy confidence intervals and 1D $\Delta\chi^2$ as a function of δ_{CP} between this analysis and other T2K analyses, from an Asimov fit with the reactor constraint on $\sin^2 2\theta_{13}$

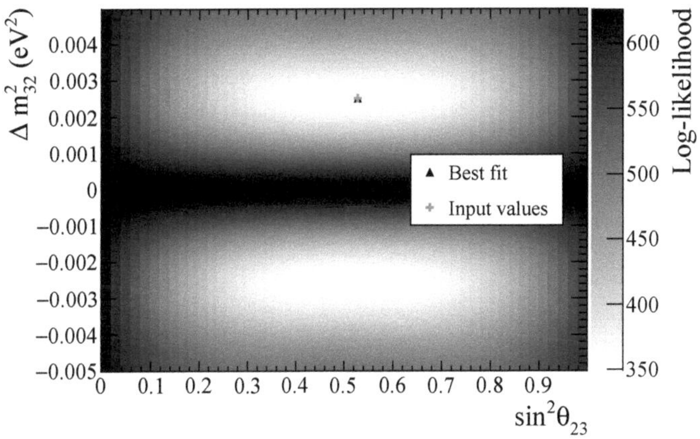

Fig. 5.19 LLH scan in $\sin^2\theta_{23}$–Δm^2_{32}, with input values $\sin^2\theta_{23}=0.528$ and $\Delta m^2_{32}=2.509\times10^{-3}\text{eV}^2$. The point with minimum LLH is the same as the input values, as expected

$\sin^2\theta_{23}$–Δm^2_{32} or $\sin^2\theta_{13}$–δ_{CP}. For each point in the grid the oscillation parameters in the MC were set to the values of either $\sin^2\theta_{23}$ and Δm^2_{32} or $\sin^2\theta_{13}$ and δ_{CP} at that point, and the log of the likelihood function (LLH) was evaluated for this Asimov fake data and MC combination. The result over the whole grid space was then plotted in a 2D histogram.

The result is shown in Figs. 5.19 and 5.20 (in $\sin^2\theta_{23}$–Δm^2_{32} and $\sin^2\theta_{13}$–δ_{CP}, respectively). The shaded 2D histogram shows the value of the LLH in each bin. A grey cross is plotted at the input values of the oscillation parameters, and a black triangle shows the minimum bin (labelled 'Best fit'). Because the fit maximises the negative of the function plotted, the bin with the maximum posterior probability will be the minimum bin here. The input and minimum LLH points are plotted in identical places in both figures, which shows that the machinery is able to correctly identify the best-fit point, and is unbiased with respect to the oscillation parameters $\sin^2\theta_{23}$, $\sin^2\theta_{13}$, Δm^2_{32}, and δ_{CP}.

Further validations of the fitter have been completed, which, although space does not permit a full discussion here, are documented in [1]. These include Asimov fits with input oscillation parameters corresponding to non-maximal mixing ($\sin^2\theta_{23}=0.45$) and no CP violation ($\delta_{CP}=0$), the results of which have been tested against the VaLOR and P-Theta oscillation groups on T2K and excellent agreement seen between all three groups. In all validation tests, we have seen that the fitter finds the input values well, and where marginalisation effects mean that it does not, these have been investigated and understood. Taken together with the tests shown in this section, these validations give us confidence that the fitter is working correctly.

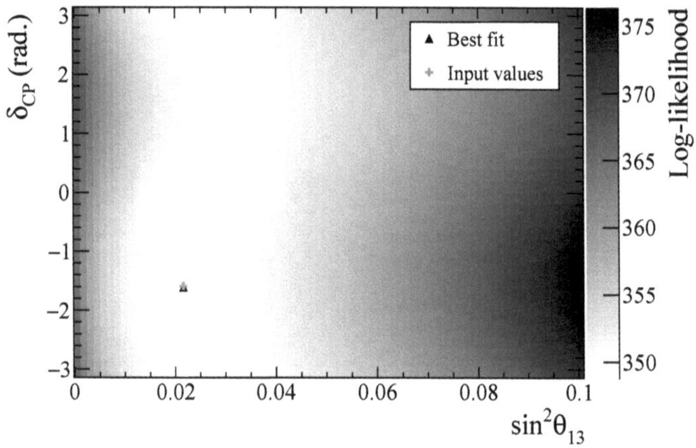

Fig. 5.20 LLH scan in $\sin^2\theta_{13}$–δ_{CP}, with input values $\sin^2\theta_{13} = 0.0217$ and $\delta_{CP} = -1.601$. This is a plot of the likelihood so does not include the prior term from the reactor constraint. The point with minimum LLH is the same as the input values, as expected

References

1. K.E. Duffy, A. Kaboth, A joint ND280-SK 1R$_\mu$-SK 1R$_e$ fit of neutrino and antineutrino-mode data using MCMC, Technical Report 269, T2K (2016), http://www.t2k.org/docs/technotes/269
2. K.A. Olive et al., (Particle Data Group), Review of particle physics, Chin. Phys. C **38**, 090001 ((2014). (and 2015 update). doi:10.1088/1674-1137/38/9/090001
3. P. Bartet et al., ν_μ CC event selections in the ND280 tracker using run 2+3+4 Data, Technical Report 212, T2K (2015), http://www.t2k.org/docs/technotes/212
4. V. Berardi et al., CC $\bar{\nu}_\mu$ event selection in the ND280 tracker using run 5c anti-neutrino beam data, Technical Report 224, T2K (2015), http://www.t2k.org/docs/technotes/224
5. V. Berardi et al., ND280 tracker analysis of ν_μ contamination in antinu beam in run5c data, Technical Report 227, T2K (2015), http://www.t2k.org/docs/technotes/227
6. V. Berardi et al., CC $\bar{\nu}_\mu$ event selection in the ND280 tracker using run 5c and run 6 anti-neutrino beam data, Technical Report 246, T2K (2015), http://www.t2k.org/docs/technotes/246
7. V. Berardi et al., CC ν_μ background event selection in the ND280 tracker using run 5c+run 6 anti-neutrino beam data, Technical Report 248, T2K (2015), http://www.t2k.org/docs/technotes/248
8. M. Hartz et al., Constraining the flux and cross section models with data from the ND280 detector using FGD1 and FGD2 for the 2016 joint oscillation analysis, Technical Report 230, T2K (2016), http://www.t2k.org/docs/technotes/230
9. J. Imber et al., Super-kamiokande events and data quality studies for T2K run 7, Technical Report 284, T2K (2016), http://www.t2k.org/docs/technotes/284
10. K. Abe et al., (T2K Collaboration), Measurements of neutrino oscillation in appearance and disappearance channels by the T2K experiment with 6.6×10^{20} protons on target. Phys. Rev. D **91**, 072010 (2015). doi:10.1103/PhysRevD.91.072010
11. M. Friend et al., Flux prediction and uncertainty updates with NA61 2009 thin target data and negative focussing mode predictions, Technical Report 217, T2K (2016), http://www.t2k.org/docs/technotes/217

12. M. Hartz et al., Constraining the flux and cross section models with data from the ND280 detector for the 2014/15 oscillation analysis, Technical Report 220, T2K (2015), http://www.t2k.org/docs/technotes/220
13. C. Wilkinson et al., Testing charged current quasi-elastic and multinucleon interaction models in the NEUT neutrino interaction generator with published datasets from the MiniBooNE and MINERνA experiments. Phys. Rev. D **93**, 072010 (2016). doi:10.1103/PhysRevD.93.072010
14. C. Wilkinson et al., Reanalysis of bubble chamber measurements of muon-neutrino induced single pion production. Phys. Rev. D **90**, 112017 (2014). doi:10.1103/PhysRevD.90.112017
15. A. Bercellie et al., Cross section parameters for 2014 oscillation analysis, Technical Report 192, T2K (2014), http://www.t2k.org/docs/technotes/192
16. A. Bercellie et al., Implementation of additional NIWG cross section parameterizations for 2014 analyses, Technical Report 193, T2K (2014), http://www.t2k.org/docs/technotes/193
17. M. Dunkman et al., Updated recommendation of the 2014-5 oscillation parameters, Technical Report 265, T2K (2016), http://www.t2k.org/docs/technotes/265
18. J. Nieves, I.R. Simo, M.J.V. Vacas, Inclusive charged-current neutrino-nucleus reactions. Phys. Rev. C **83**, 045501 (2011). doi:10.1103/PhysRevC.83.045501
19. R. Gran, J. Nieves, F. Sanchez, M.J.V. Vacas, Neutrino-nucleus quasi-elastic and 2p2h interactions up to 10 GeV. Phys. Rev. D **88**, 113007 (2013). doi:10.1103/PhysRevD.88.113007
20. M. Martini, M. Ericson, G. Chanfray, J. Marteau, Unified approach for nucleon knock-out and coherent and incoherent pion production in neutrino interactions with nuclei. Phys. Rev. C **80**, 065501 (2009). doi:10.1103/PhysRevC.80.065501
21. Melanie Day, Kevin S. McFarland, Differences in quasielastic cross sections of muon and electron neutrinos. Phys. Rev. D **86**, 053003 (2012). doi:10.1103/PhysRevD.86.053003
22. A. Higuera et al., (MINERvA Collaboration), Measurement of coherent production of $\pi^{\pm}$ in neutrino and antineutrino beams on carbon from E_{ν} of 1.5 to 20 GeV. Phys. Rev. Lett. **113**, 261802 (2014). doi:10.1103/PhysRevLett.113.261802
23. M. Hartz, A. Kaboth, K. Mahn, Constraining the flux and cross section models with data from the ND280 detector for the 2013 oscillation analysis, Technical Report 166, T2K (2013), http://www.t2k.org/docs/technotes/166
24. P. de Perio, J. Imber, Super-K systematic uncertainties for run 1-4 Joint ν_e+ν_{μ} analyses, Technical Report 186, T2K (2014), http://www.t2k.org/docs/technotes/186
25. P. de Perio et al., Cross section parameters for the 2012a oscillation analysis, Technical Report 108, T2K (2012), http://www.t2k.org/docs/technotes/108
26. T2K Beam Group, flux releases summary, Technical Report 264, T2K (2016), http://www.t2k.org/docs/technotes/264
27. E. Wang et al., Photon emission in neutral current interactions at the T2K experiment. Phys. Rev. D **92**, 053005 (2015). doi:10.1103/PhysRevD.92.053005
28. C. Andreopoulos et al., T2K neutrino and anti-neutrion 3-flavour joint analysis of run 1–6 (6.914×10^{20}-POT ν 4.011×10^{20}-POT $\bar{\nu}$) and run 1–7 (7.002×10^{20}-POT ν 7.471×10^{20}-POT $\bar{\nu}$) data sets, Technical Report 266, T2K (2016), http://www.t2k.org/docs/technotes/266
29. J. Imber for the P-theta group, four sample joint oscillation analysis with T2K run 1–7 data, Technical Report TN-267, T2K (2016), http://www.t2k.org/docs/technotes/267

Chapter 6
Joint ν+$\bar{\nu}$ Oscillation Analysis: Results

This section details the results of fitting simultaneously the T2K ND280 and SK neutrino and antineutrino-mode data. The data samples were fitted twice: once with a flat prior on $\sin^2\theta_{13}$ (the 'T2K-only' fit), and once using a Gaussian prior on $\sin^2 2\theta_{13}$ according to the measurements by reactor experiments reported in the 2015 edition of the PDG: $\sin^2 2\theta_{13} = 0.085 \pm 0.005$ (the 'T2K+Reactor' fit). In both cases, a Markov chain of 1.5×10^7 steps after burn-in was used for the fit.

Figure 6.1 shows the posterior distribution from both fits in 2D $\sin^2\theta_{23}$–Δm^2_{32} parameter space. The mode of the 2D and 4D marginal posterior distributions are very similar in both $\sin^2\theta_{23}$ and Δm^2_{32}, which implies that there is no strong marginalisation effect from $\sin^2\theta_{13}$ or δ_{CP}. However, we do see a hint of the marginalisation effects seen in Sects. 5.5.1.1 and 5.5.1.2: the mode of the 2D posterior is slightly lower in $\sin^2\theta_{23}$ than the mode of the 4D posterior for the T2K-only fit, and slightly higher for the T2K+Reactor fit.

Comparing the two figures, we see that both are largely similar (indeed, both have the same best-fit point in the 4D marginal posterior, as shown in Table 6.1). The effect of the reactor constraint seems to be to push the posterior probability towards higher $\sin^2\theta_{23}$. The portion of the credible interval contours in inverted hierarchy is also smaller for the T2K+Reactor fit, implying that the normal hierarchy is more strongly favoured. This is even more obvious in the 1D marginal posterior as a function of Δm^2_{32}, shown in Fig. 6.2, where the 1D 68% credible interval for the T2K+Reactor fit excludes the inverted hierarchy entirely. The effect of marginalisation over the mass hierarchy in the T2K+Reactor fit can also be seen in the 1D marginal posterior distribution as a function of $\sin^2\theta_{23}$, shown in Fig. 6.3. The upper octant is mildly favoured in the 2D marginal posterior in both mass hierarchies, and that tendency is exacerbated in the 1D marginal posterior. The result of marginalising over the mass hierarchy is a very asymmetric probability distribution, and an asymmetric uncertainty on the best-fit value of $\sin^2\theta_{23}$ given in Table 6.1.

K.E. Duffy, *First Measurement of Neutrino and Antineutrino Oscillation at T2K*, Springer Theses, DOI 10.1007/978-3-319-65040-1_6

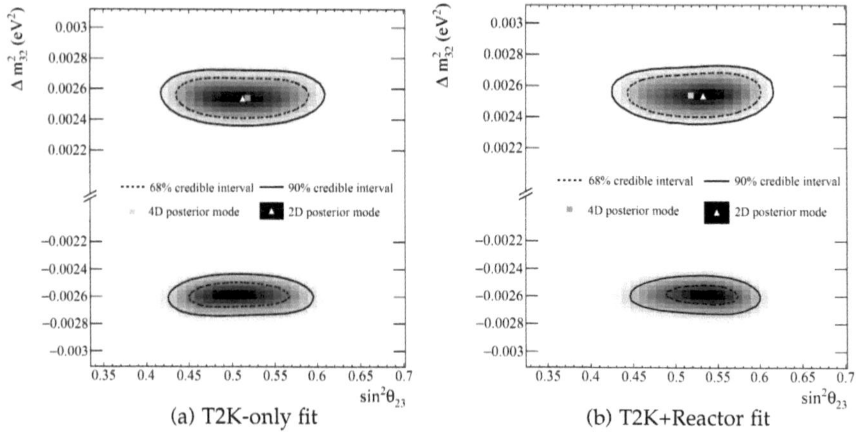

Fig. 6.1 2D posterior and credible intervals in $\sin^2\theta_{23}$–Δm^2_{32} from both the T2K-only and T2K+Reactor data fit

Table 6.1 Best-fit values and uncertainties from the T2K-only and T2K+Reactor data fits. The best-fit values are taken from the mode of the 4D marginal posterior probability distribution in $\sin^2\theta_{23}$–Δm^2_{32}–$\sin^2\theta_{13}$–δ_{CP}, and the uncertainty on each parameter is taken from the 68% credible interval of the 1D marginal posterior probability distribution. No uncertainties are given for δ_{CP} because the distribution is non-Gaussian and cyclical

	Δm^2_{32} ($\times 10^{-3}$eV^2)	$\sin^2\theta_{23}$	$\sin^2\theta_{13}$	δ_{CP} (rad.)
T2K-only	$2.54^{+0.12}_{-0.10}$	$0.519^{+0.031}_{-0.059}$	$0.0257^{+0.0106}_{-0.0440}$	-1.91
T2K+Reactor	$2.54^{+0.12}_{-0.12}$	$0.519^{+0.061}_{-0.029}$	$0.0219^{+0.0014}_{-0.0014}$	-1.72

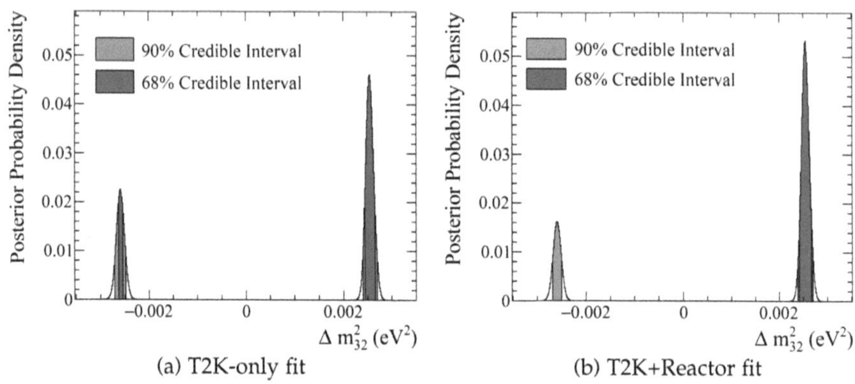

Fig. 6.2 1D posterior and credible intervals in Δm^2_{32} from both the T2K-only and T2K+Reactor data fit

The 2D marginal posterior distributions in $\sin^2\theta_{13}$–δ_{CP} space are shown in Fig. 6.4. The T2K-only result is particularly interesting: with the new antineutrino-mode data we now see a 90% closed contour in $\sin^2\theta_{13}$–δ_{CP} (recall that previous T2K

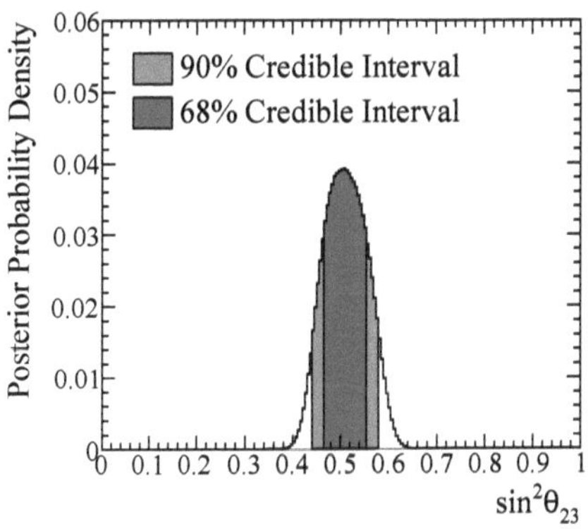

(a) T2K-only fit

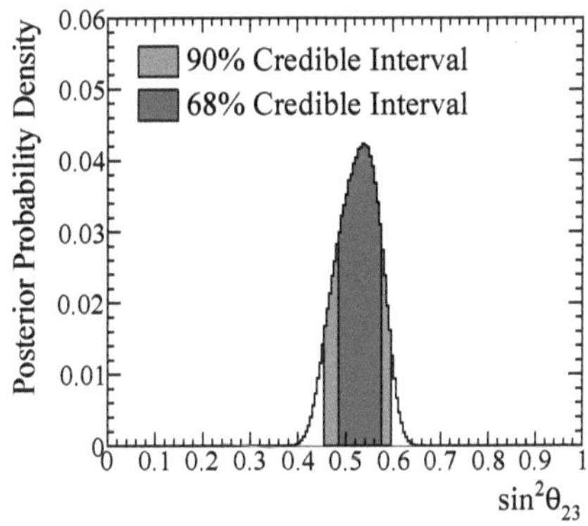

(b) T2K+Reactor fit

Fig. 6.3 1D posterior and credible intervals in $\sin^2\theta_{23}$ from both the T2K-only and T2K+Reactor data fit

fits with neutrino-mode data only—shown in Sect. 6.2—had no sensitivity to δ_{CP} and could not produce any closed contour in this space). As in $\sin^2\theta_{23}$–Δm^2_{32} space, we see good agreement between the modes of the 2D and 4D marginal posterior probability distributions, in both fits. There is a small difference between the two points in δ_{CP} for the T2K-only fit, but the posterior distribution is very flat, meaning that the difference in posterior probability between these two points is small.

The effect of the reactor constraint is obvious when comparing Fig. 6.4a, b, as well as in Fig. 6.5, which plots the 1D marginal posterior probability density as a function of $\sin^2\theta_{13}$. The T2K data alone prefer a higher value of $\sin^2\theta_{13}$, with a fairly wide posterior probability distribution. However, the T2K data do not have much power to 'pull' against the prior given by the reactor constraint, so the T2K+Reactor fit is almost entirely dominated (in $\sin^2\theta_{13}$) by the reactor measurement.

In both the T2K-only fit and the T2K+Reactor fit, the distribution peaks around $\delta_{CP}\simeq -\pi/2$, with a more sharply-peaked distribution (shown by smaller credible interval contours) for the T2K+Reactor fit. This is demonstrated even more clearly in Fig. 6.6, in which the marginal posterior probability distributions as a function of δ_{CP} are presented. It is gratifying to note that both T2K data alone and T2K data combined with the reactor constraint favour the same maximal value of δ_{CP}.

Table 6.1 shows the best-fit values and 68% uncertainty for each of the interesting oscillation parameters, from both the T2K-only and the T2K+Reactor fits. The best-fit values are extracted from the mode of the 4D posterior, and the uncertainties are taken from the 68% credible interval of the 1D marginal posterior as a function of that parameter (as shown in Figs. 6.2, 6.3, 6.5, and 6.6). No uncertainty is given on the parameter estimate for δ_{CP} because the distribution is non-Gaussian and cyclical, so to quote an uncertainty in this way might be misleading. Instead, the 68, 90, and 95% 1D credible interval ranges for δ_{CP} are shown in Table 6.2.

Table 6.3 shows the posterior probability (given the data and prior assumptions) that a given combination of mass hierarchy and octant is correct. Both fits favour

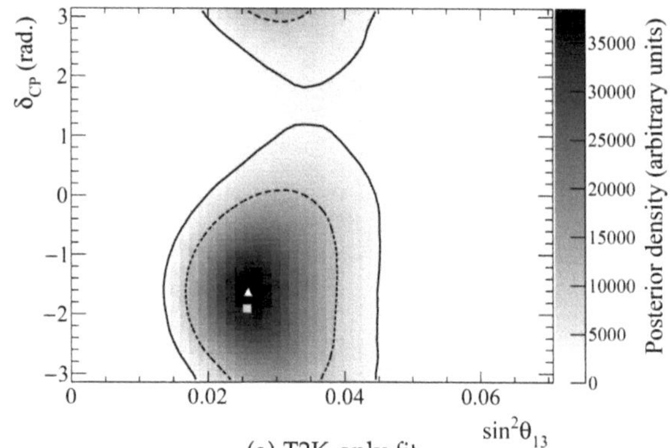

(a) T2K-only fit

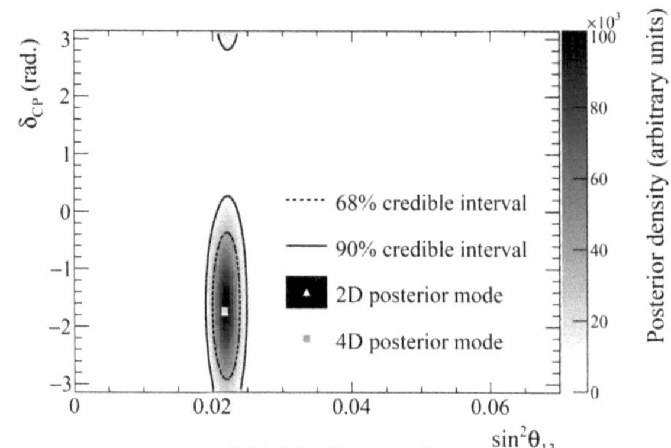

(b) T2K+Reactor fit

Fig. 6.4 2D posterior and credible intervals in $\sin^2\theta_{13}$–δ_{CP} from both the T2K-only and T2K+Reactor data fit

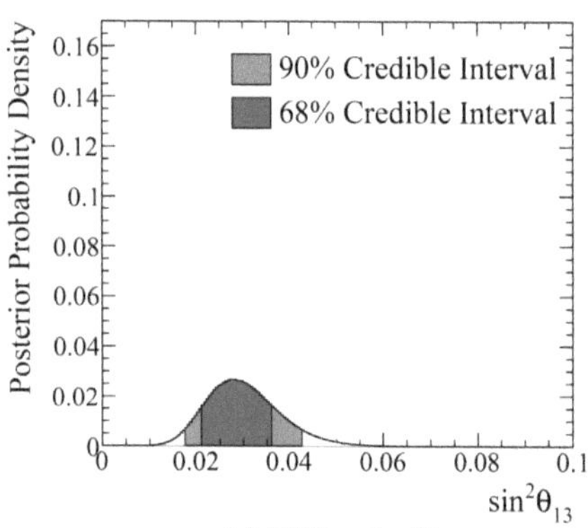

(a) T2K-only fit

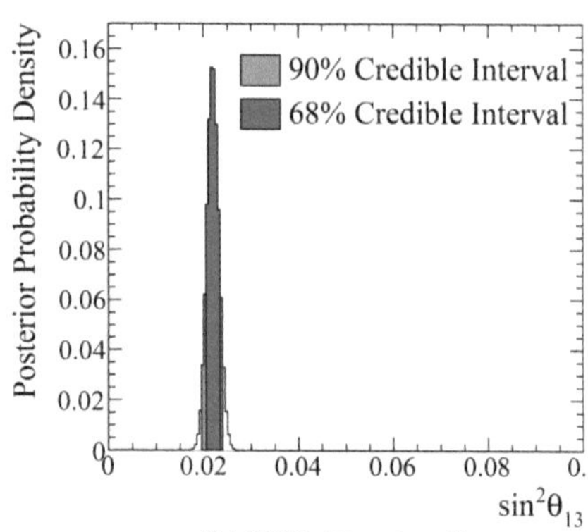

(b) T2K+Reactor fit

Fig. 6.5 1D posterior and credible intervals in $\sin^2\theta_{13}$ from both the T2K-only and T2K+Reactor data fit

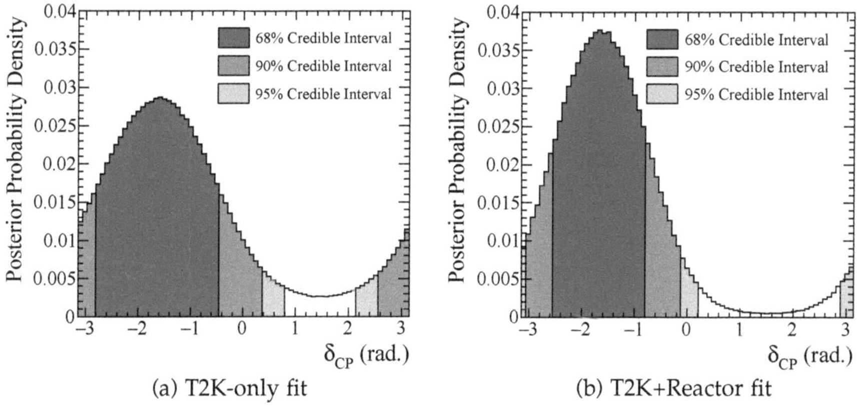

Fig. 6.6 1D posterior and credible intervals in δ_{CP} from both the T2K-only and T2K+Reactor data fit

Table 6.2 1D 68, 90, and 95% included credible interval ranges for δ_{CP} (rad.) on a scale of $-\pi$–π, from both the T2K-only and T2K+Reactor fits

	68% C.I. range	90% C.I. range	95% C.I. range
T2K-only	$-2.81 - -0.461$	$-\pi - 0.377$ & $2.60 - \pi$	$-\pi - 0.796$ & $2.18 - \pi$
T2K+Reactor	$-2.51 - -0.754$	$-3.10 - -0.168$	$-\pi - 0.168$ & $2.89 - \pi$

Table 6.3 Model comparison probabilities for normal and inverted hierarchies as well as upper and lower octants, from the posterior of the T2K-only fit and the T2K+Reactor fit

	$\sin^2\theta_{23} < 0.5$	$\sin^2\theta_{23} > 0.5$	Sum
T2K-only			
IH ($\Delta m^2_{32} < 0$)	0.15	0.29	0.34
NH ($\Delta m^2_{32} > 0$)	0.27	0.39	0.66
Sum	0.42	0.58	1
T2K+Reactor			
IH ($\Delta m^2_{32} < 0$)	0.07	0.18	0.25
NH ($\Delta m^2_{32} > 0$)	0.22	0.53	0.75
Sum	0.29	0.71	1

the normal mass hierarchy and upper octant, although both are favoured with more significance in the T2K+Reactor fit. However, in neither case is the result statistically significant. For the T2K-only fit, the mass hierarchy Bayes factor B(NH/IH) = 1.94 and the octant Bayes factor, B($\sin^2\theta_{23}$> 0.5/$\sin^2\theta_{23}$< 0.5), is 1.38. For the T2K+Reactor fit, the mass hierarchy Bayes factor B(NH/IH) = 3.00 and the octant Bayes factor B($\sin^2\theta_{23}$> 0.5/$\sin^2\theta_{23}$< 0.5) = 2.45. These Bayes factors are all small. The most significant conclusion we can draw—from the T2K+Reactor fit—is that

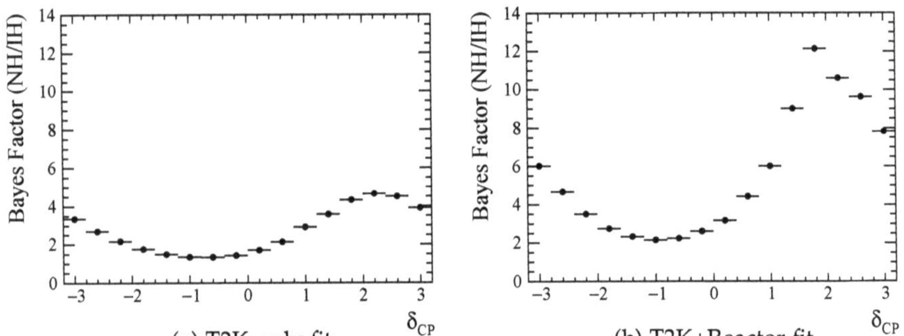

(a) T2K-only fit (b) T2K+Reactor fit

Fig. 6.7 Bayes factor for normal over inverted mass hierarchy as a function of δ_{CP}, from the posterior of the T2K-only and T2K+Reactor fit. Error bars show statistical uncertainties from the number of steps in the MCMC for a given δ_{CP} bin. We find the overall Bayes factor B(NH/IH) = 1.94 for the T2K-only fit and 3.00 for the T2K+Reactor fit when considering all points in the posterior distribution, weighted according to their posterior probabilities

the data predict that the normal mass hierarchy is three times more likely to be the correct model than the inverted mass hierarchy, and the upper octant is around 2.5 times more likely than the lower octant. This does not constitute strong evidence.

Figure 6.7 shows the Bayes factor for normal over inverted mass hierarchy, B(NH/IH), as a function of δ_{CP}. This plot is constructed by binning the posterior probability distribution in bins of δ_{CP}, and then for each bin calculating the mass hierarchy Bayes factor. It shows that the significance at which the normal mass hierarchy is favoured depends on the value of δ_{CP}—although the normal mass hierarchy is favoured for all values of δ_{CP}, the significance peaks around $\delta_{CP} \simeq 2$ and is minimal around $\delta_{CP} \simeq -1$. When comparing this figure to the 'overall' Bayes factors derived from Table 6.3, it is important to remember that the Bayes factor calculated at each point in Fig. 6.7 must be weighted according to the posterior probability for that value of δ_{CP}. It may be counterintuitive to think that the 'overall' Bayes factor corresponding to Fig. 6.7b is 3, but this can be understood by considering that not all points on the figure have equal weight. The points around $\delta_{CP} = -\pi/2$ have significantly larger posterior probabilities than points at other values of δ_{CP}, and therefore contribute much more strongly to the overall Bayes factor.

This can perhaps be understood by considering how the predicted number of events in the $1R_e$ samples changes for different values of δ_{CP} and the mass hierarchy, as shown in Table 6.4. We see that the observed number of events is lower than predicted in the neutrino-mode $1R_e$ sample and higher than predicted in the antineutrino-mode $1R_e$ sample, and the prediction which is most consistent with the measured data is generated with $\delta_{CP} = -\pi/2$ and normal mass hierarchy. In general the normal-hierarchy predictions give higher numbers of events in neutrino mode and lower numbers of events in antineutrino mode, which is consistent with the fact that the data prefer normal hierarchy over the whole δ_{CP} space. The 'worst' possible prediction (least consistent with the data) is at $\delta_{CP} = \pi/2$ and inverted mass hierarchy, which

Table 6.4 Predicted number of events in the neutrino-mode and antineutrino-mode 1R_e samples for a range of values of δ_{CP}. Other oscillation parameters were taken from previous T2K fit results and the 2015 PDG: $\sin^2\theta_{12}=0.304$, $\sin^2\theta_{13}=0.0217$, $\Delta m^2_{21}=7.53\times10^{-5}\text{eV}^2$, $\sin^2\theta_{23}=0.528$, $\Delta m^2_{32}=\pm2.509\times10^{-3}\text{eV}^2$ (normal/inverted hierarchy)

Mass hierarchy	ν-mode 1R_e		$\bar{\nu}$-mode 1R_e	
	Normal	Inverted	Normal	Inverted
$\delta_{CP}=-\pi/2$	27.0	23.9	6.0	6.5
$\delta_{CP}=0$	22.7	19.9	6.9	7.4
$\delta_{CP}=\pi/2$	18.5	16.1	7.7	8.4
$\delta_{CP}=\pm\pi$	22.7	20.2	6.8	7.4
Data	32		4	

may explain why the inverted hierarchy is disfavoured most strongly at this value of δ_{CP}. We do not see a one-to-one mapping between the predictions in this Table 6.3 and Fig. 6.7 because in the fit the other oscillation parameters are not fixed: they can move around and there will be some marginalisation effect when constructing these plots (indeed, this is why we see smaller Bayes factors for the T2K-only fit—$\sin^2\theta_{13}$ has more freedom to move to higher values and increase the event rate in the 1R_e samples, thus compensating for the effect of δ_{CP} and the mass hierarchy in the neutrino-mode sample). However, these predictions still give a useful insight into the results shown in Fig. 6.7.

Figure 6.8 shows the reconstructed energy spectra obtained using the posterior predictive method described in Sect. 4.4.5 for each of the four SK samples after the T2K+Reactor fit. The result of sampling the fit posterior 2500 times is shown as the shaded histogram (z-axis units are arbitrary, but are proportional to the marginal posterior density), and the resulting best-fit spectrum is shown with a black dashed line. The data points are overlaid for comparison. The 1R_μ samples are shown for a reduced energy range, up to 7 GeV, because no data are found at higher energies (although the fit region extends to 30 GeV).

A goodness-of-fit p-value was also calculated for each fit, using the same posterior predictive method, as described in Sect. 4.4.6. This was found to be 0.54 for the T2K-only fit and 0.51 for the T2K+Reactor fit, indicating no disagreement between either fit and the data.

Figure 6.9 shows the best-fit reconstructed energy spectra after the T2K-only fit and after the T2K+Reactor fit, along with the expectation in the absence of neutrino oscillations. Figure 6.9a, c show the characteristic 'oscillation dip' in the ratio of the best-fit spectra and data to the unoscillated prediction, which is clear evidence of ν_μ or $\bar{\nu}_\mu$ disappearance. Figure 6.9b shows a significant excess in the measured data in the neutrino-mode 1R_e sample over the unoscillated prediction (as in the first T2K measurement of ν_e appearance [1]), but the same effect is not so obvious in Fig. 6.9d, most likely due to the current low statistics in the RHC 1R_e sample. The effect of the reactor constraint, in restricting $\sin^2\theta_{13}$ to low values, is easy to see; it reduces

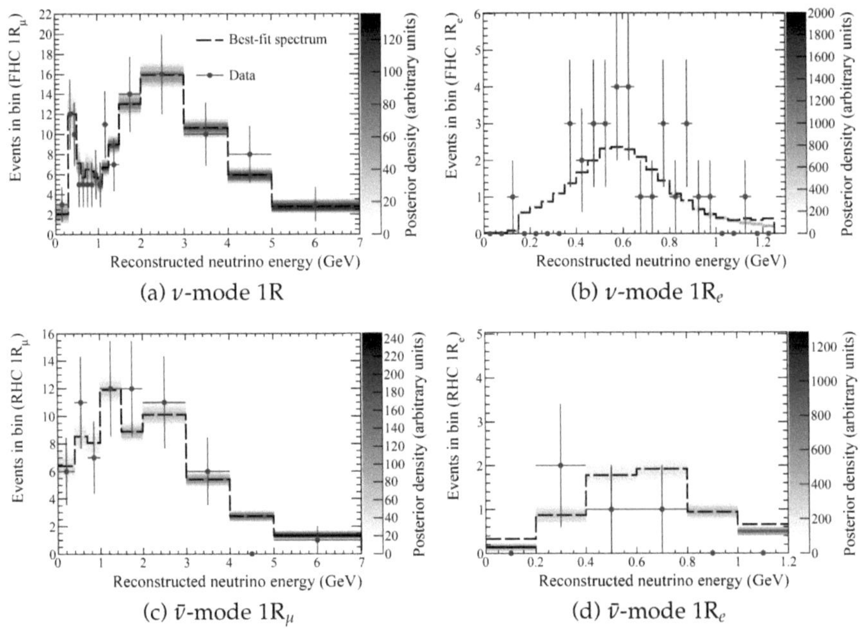

(a) ν-mode 1R (b) ν-mode 1R$_e$

(c) $\bar{\nu}$-mode 1R$_\mu$ (d) $\bar{\nu}$-mode 1R$_e$

Fig. 6.8 Result of the posterior predictive method for each of the four SK samples after the T2K+Reactor fit. Data points are shown in *black*, the best-fit reconstructed energy spectrum is shown as a *black dashed line*, and the result of sampling the fit posterior 2500 times is shown as the shaded 2D histogram. A coarser binning is used than in the data fit to reduce the effect of statistical fluctuations and make the shape of the result easier to see

the predicted number of events in the 1R$_e$ samples. There is no noticeable effect on the 1R$_\mu$ samples because the ν_μ disappearance probability depends only weakly on $\sin^2\theta_{13}$.

6.1 Systematic Parameters

The aim of this analysis is to estimate the values of the oscillation parameters $\sin^2\theta_{13}$, $\sin^2\theta_{23}$, Δm^2_{32}, and δ_{CP}. However, it is instructive to also examine how the systematic parameters described in Sect. 5.3 behave in the fit, because the distributions of these parameters affect the marginal posterior distribution of the interesting oscillation parameters.

Figures 6.10, 6.11, 6.12, and 6.13 show the central values and 1D 1σ ranges for each of the systematic parameters used in the fit. The light grey points and shaded ranges show the prior central values and $\pm 1\sigma$ ranges, the black points and lines show the mode and $\pm 1\sigma$ ranges after the T2K+Reactor fit, and the grey points and lines show the mode and $\pm 1\sigma$ ranges after the T2K-only fit (in both cases found by fitting

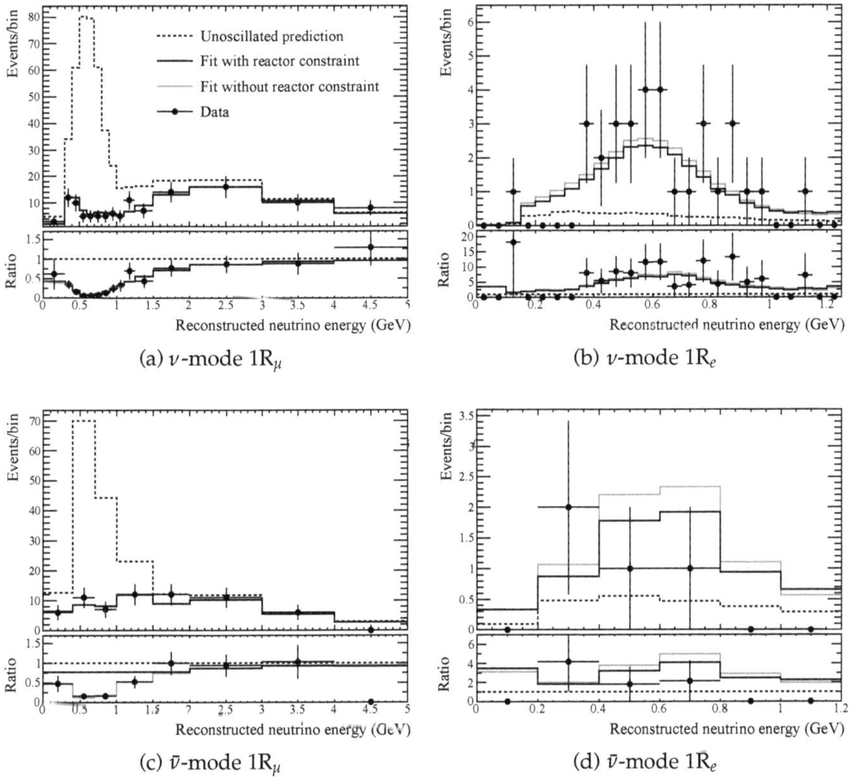

Fig. 6.9 Best-fit energy spectra and ratio to unoscillated prediction for each of the four SK samples. A coarser binning is used than in the data fit to reduce the effect of statistical fluctuations and make the shape of the result easier to see, and the $1R_\mu$ spectra are shown for low energies only to highlight the oscillation dip around 0.6 GeV

a Gaussian around the peak of the 1D posterior distribution for each parameter). The ‘allowed’ ranges are also shown for those cross-section parameters which are only valid in reduced ranges.

In general we see good agreement in the systematics between the T2K-only and T2K+Reactor fits, as we might expect—the ND280 data samples, which dominate the fit for the systematic parameters, are identical. Differences between the two fits are small, and are understood to be due to correlations with the other parameters, particularly the oscillation parameters, which have different distributions in the T2K-only and T2K+Reactor fits. The flux parameters are increased in both fits by around 1σ from their prior values, driven by the pre-fit deficit in the MC prediction for the ν_μ CC 0π and ν_μ CC Other samples, but most other parameters are consistent with the prior estimations. All of the detector systematic parameters and most of the cross-section parameters have central values that are within 1σ (according to the prior uncertainty) of the central point of the prior. The uncertainty on the ND280

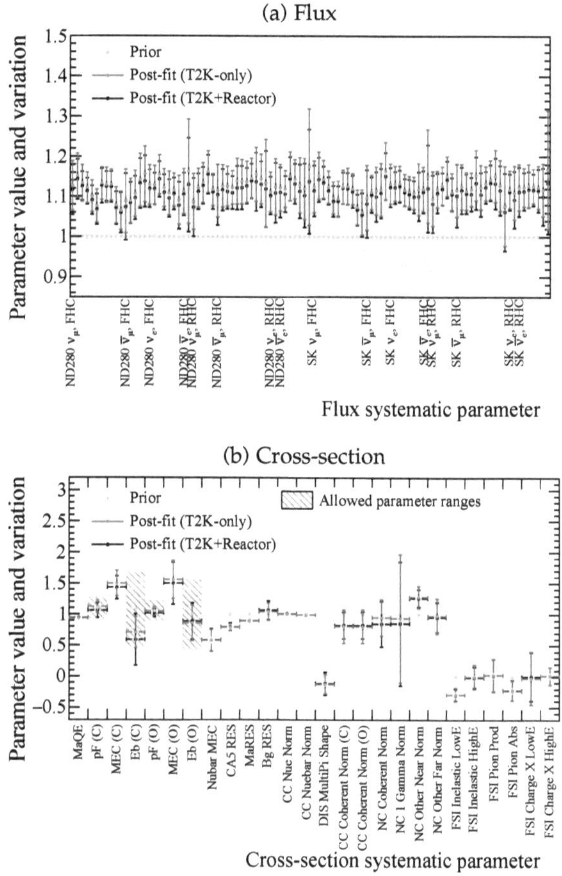

Fig. 6.10 Central values and 1σ uncertainties for the flux and cross-section systematic parameters, before and after the data fit. No 'prior' central values or uncertainties are shown for those cross-section systematic parameters for which flat priors are used. The 'allowed' parameter ranges for those parameters which have reduced ranges are also shown

detector and cross-section systematic parameters is reduced by the fit, indicating that the data provide a noticeably stronger constraint on the parameters than the prior. Only the SK detector systematic parameters are relatively unchanged by the data fit, with central values and uncertainties close to the prior, most likely because the statistics in the SK data samples are much smaller than at ND280.

6.2 Comparison to Previous T2K Results

Although this is the first T2K analysis to fit neutrino and antineutrino data, it is not the first T2K joint fit of ν_μ disappearance and ν_e appearance. A previous analysis [2], already mentioned in this thesis, fit both $1R_\mu$ and $1R_e$ samples from the neutrino-mode data in T2K runs 1-4, with a total of 6.6×10^{20} POT.

Figures 6.14 and 6.15 show the comparison of credible intervals from the run 1-4 data fit with those presented in this thesis, for the T2K-only fit and the T2K+Reactor

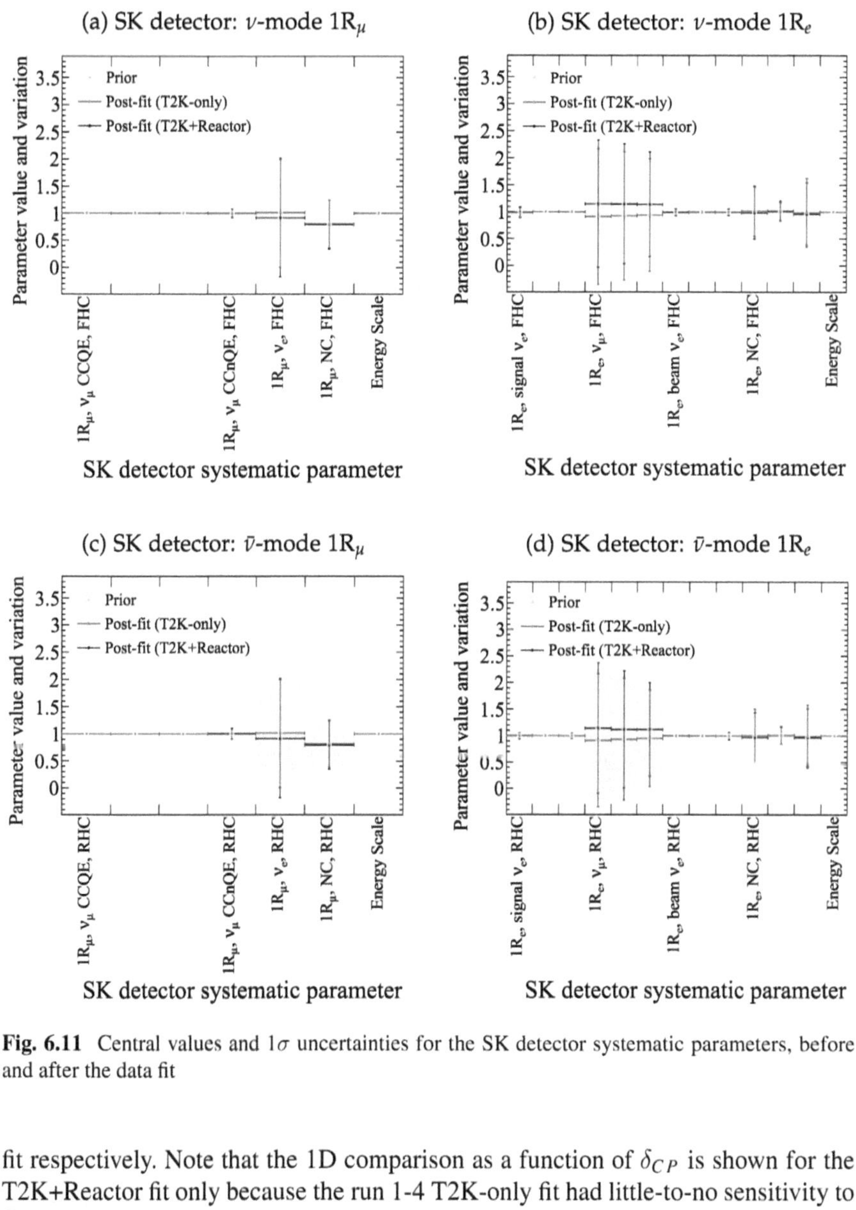

Fig. 6.11 Central values and 1σ uncertainties for the SK detector systematic parameters, before and after the data fit

fit respectively. Note that the 1D comparison as a function of δ_{CP} is shown for the T2K+Reactor fit only because the run 1-4 T2K-only fit had little-to-no sensitivity to δ_{CP}, so this plot was not produced.

In both cases the $\sin^2\theta_{23}$–Δm^2_{32} contours are consistent with the previous result, but the size of the contours has decreased, meaning that the additional data has improved the precision of these measurements. In general the contours are slightly shifted to larger values of $|\Delta m^2_{32}|$, which is also reflected in the movement of the best-fit point. The normal mass hierarchy is now favoured both in the T2K-only and T2K+Reactor fit, whereas the inverted hierarchy was favoured in the T2K-only

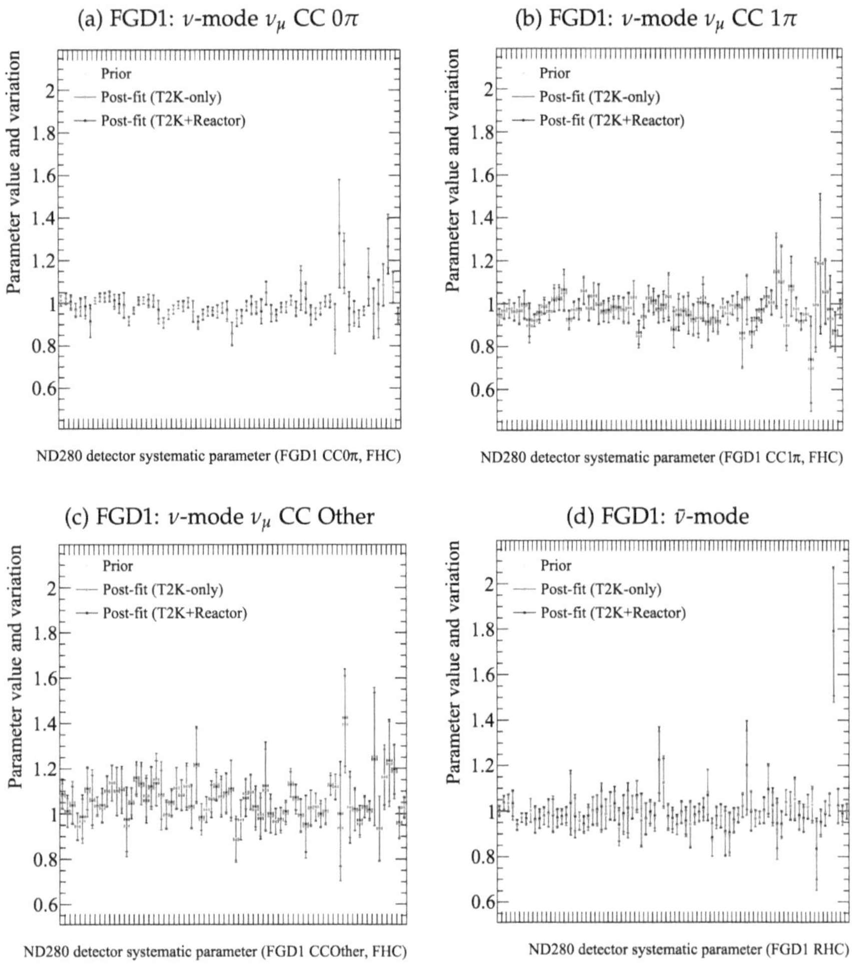

Fig. 6.12 Central values and 1σ uncertainties for the ND280 FGD1 detector systematic parameters, before and after the data fit

run 1-4 fit. This is most likely a result of the correlation between mass hierarchy and δ_{CP} discussed above: the previous T2K-only fit had no sensitivity to δ_{CP} so the marginalisation onto $\sin^2\theta_{23}$–Δm^2_{32} will have included all values of δ_{CP} with roughly equal probabilities. By comparing the relative sizes of the contours in Fig. 6.14a it is clear that the hierarchy preference in the T2K-only run 1-4 fit was extremely weak, so the change of hierarchy preference is not significant.

As previously mentioned, the most interesting result—and the most significant change that we would expect as a result of adding antineutrino data—is the ability to constrain δ_{CP}. Whereas the previous neutrino-only fit had no power to measure δ_{CP} when fitting T2K data only, we now see a closed 68 and 90% contour

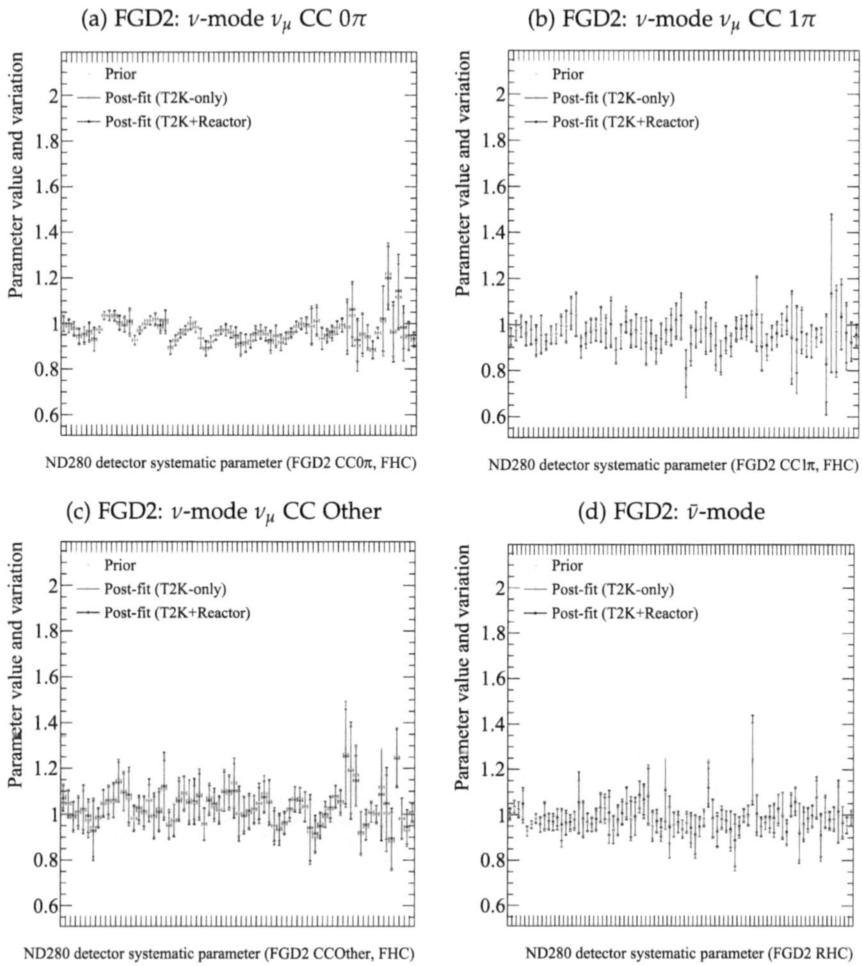

Fig. 6.13 Central values and 1σ uncertainties for the ND280 FGD2 detector systematic parameters, before and after the data fit

in $\sin^2\theta_{13}$–δ_{CP}. In the fit with the reactor constraint, the probability distribution has become even more sharply peaked in δ_{CP} (with a 90% exclusion of $\delta_{CP}=0$ and $\pm\pi$, compared to a 68% exclusion in the previous fit), but the result of this fit is consistent with the hint towards $\delta_{CP}=-\pi/2$ seen in the previous analysis. The most obvious difference between the two T2K+Reactor fit results is the change in the reactor constraint itself. The constraint used for the run 1-4 neutrino-mode fit came from the 2013 edition of the PDG [3]: $\sin^2 2\theta_{13} = 0.095 \pm 0.01$, with central value corresponding to $\sin^2\theta_{13} = 0.0243$. The reactor measurement of $\sin^2 2\theta_{13}$ has since been updated, $\sin^2 2\theta_{13} = 0.085 \pm 0.005$, which has a central value corresponding

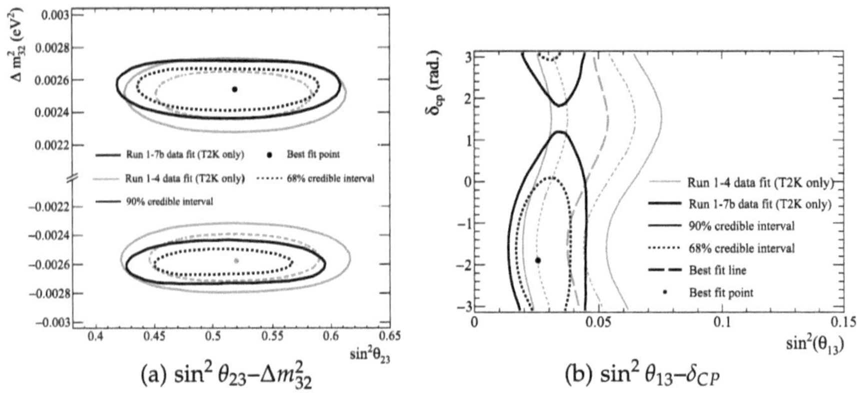

Fig. 6.14 Results of the T2K-only fit described in this thesis (labelled "Run 1-7b data") compared to the results of a fit to T2K run 1-4 data, as reported in [2]

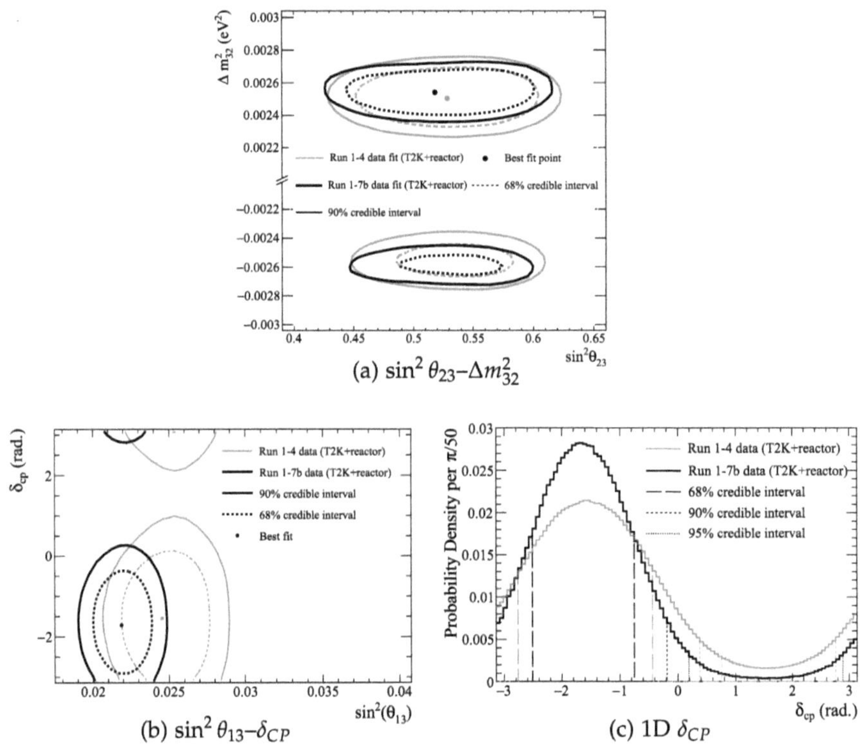

Fig. 6.15 Results of the T2K+Reactor fit described in this thesis (labelled "Run 1-7b data") compared to the results of a fit to T2K run 1-4 data, as reported in [2]

to $\sin^2\theta_{13} = 0.0217$. Both the central value and the 1σ uncertainty of the prior on $\sin^2 2\theta_{13}$ has been reduced, and this is reflected in the contours in Fig. 6.15b.

6.3 Comparison to Other Neutrino Oscillation Experiments

Figure 6.16 shows the credible intervals in $\sin^2\theta_{23}$–Δm^2_{32} from the T2K+Reactor fit compared with published results from other neutrino oscillation experiments. Measurements in the $\sin^2\theta_{23}$–Δm^2_{32} plane were included from NOνA [4], Super-Kamiokande [5], IceCube DeepCore [6], and MINOS and MINOS+ [7]. These results were not all shown in Sect. 2.2.3 because some of them are too new to have been included in the Particle Data Group review, or not competitive with the current world-leading measurements. In order to make the figure easier to read, the 68 and 90% credible intervals and best-fit point are shown for T2K, but only the 90% confidence level contours are plotted for other experiments.

When looking at Fig. 6.16 it is important to bear in mind that this is not a direct comparison: the T2K contours shown here are Bayesian credible intervals, while the contours from other experiments are frequentist confidence intervals. As discussed

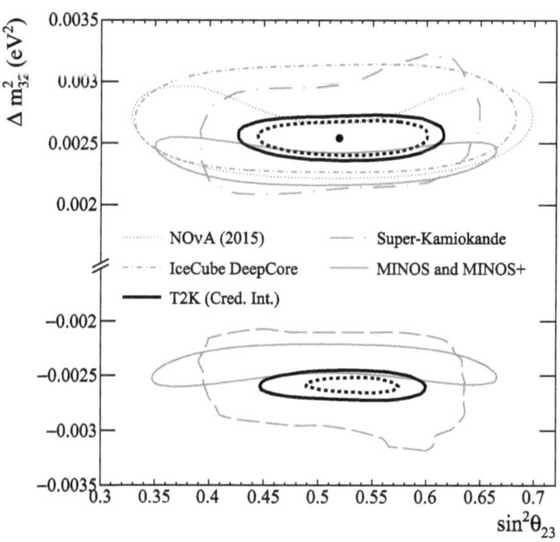

Fig. 6.16 Credible intervals in $\sin^2\theta_{23}$–Δm^2_{32} from the T2K+Reactor data fit compared with published results from other neutrino oscillation experiments: NOνA [4], Super-Kamiokande [5], IceCube DeepCore [6], and MINOS and MINOS+ [7]. Note that this is not an exact comparison because the T2K contours are Bayesian credible intervals, whereas the rest of the contours shown are frequentist confidence intervals. The 68% (*dotted line*) and 90% (*solid line*) credible intervals, as well as the best-fit point, are shown for T2K, and only the 90% confidence intervals are shown for the other experiments

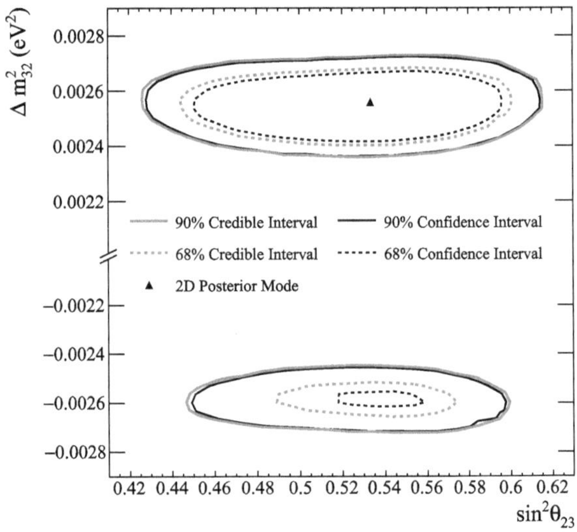

Fig. 6.17 Bayesian credible interval and frequentist confidence intervals in $\sin^2\theta_{23}$–Δm^2_{32} from the T2K+Reactor data fit. The confidence intervals are calculated using a constant-$\Delta\chi^2$ method, with respect to the overall 2D best-fit point (which is in the normal hierarchy)

in Sect. 4.4.3, these have different interpretations and meanings, and it is not always possible to compare the two. However, in this case, because the posterior distribution is close to Gaussian in both $\sin^2\theta_{23}$ and Δm^2_{32}, the credible intervals and constant $-\Delta\chi^2$ confidence intervals obtained from this fit are fairly similar (at least for the 90% contours), as shown in Fig. 6.17. In fact, the credible intervals give a slightly more conservative estimate of the uncertainty on the oscillation parameters than the confidence intervals.

6.4 Comparison to Sensitivity

As stated in Sect. 5.5.1, the Asimov fits used for validation of the fitter are indicative of the sensitivity of the experiment. An Asimov fit shows what we would measure in the absence of statistical fluctuations, so comparing the results of the data fit to the results of an Asimov fit will give an insight into the effect of statistical fluctuations in the data.

The Asimov fits presented in Sect. 5.5.1 use the best-fit points from the T2K run 1-4 fit as the input values for $\sin^2\theta_{23}$, Δm^2_{32}, and δ_{CP}, and the value given in the 2015 PDG for $\sin^2\theta_{13}$. Section 6.2 shows that the results of the data fit are largely consistent with these values, and therefore it is reasonable to compare the results of the data fits to these Asimov fits. While this is not the perfect test (it would perhaps be more correct to recreate Asimov fits using the best-fit parameters from the data fit), it is sufficient for a general comparison of the results to the expected sensitivity.

These comparisons are shown in Figs. 6.18 and 6.19 for the T2K-only and T2K+Reactor fits respectively. In almost all cases the credible intervals from the

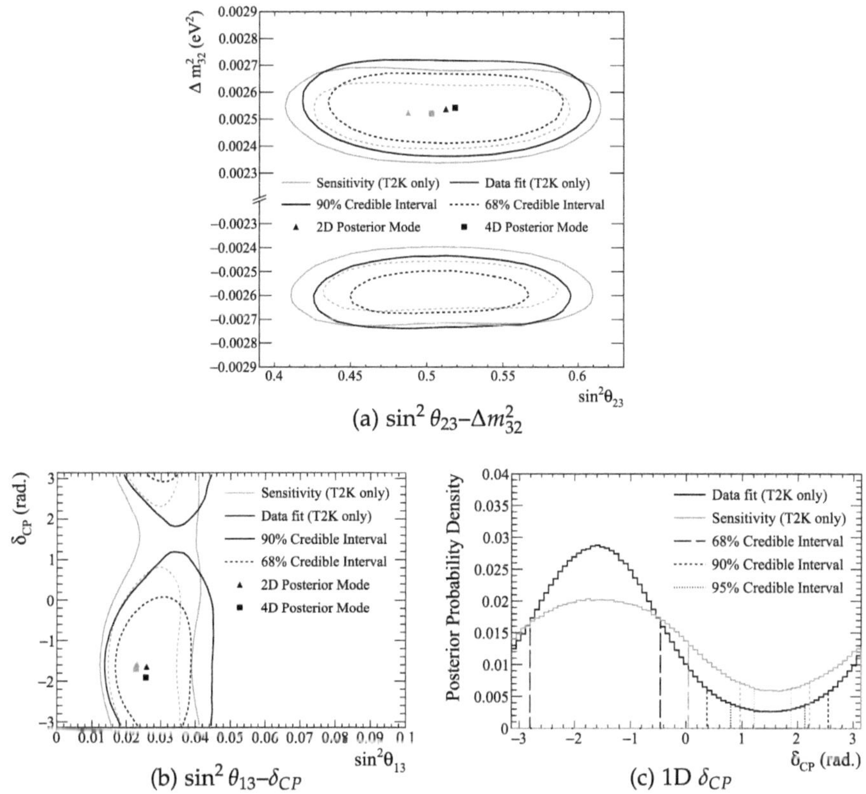

Fig. 6.18 Results of the T2K-only fit compared to the T2K-only Asimov fit presented in Sect. 5.5.1

data fit are smaller than those from the Asimov fit, showing that the fit results are 'better' than the sensitivity. The one exception to this is $\sin^2\theta_{13}$, in which the T2K-only data fit prefers a higher value of $\sin^2\theta_{13}$ than the reactor measurement (which was used as the input for the Asimov fit), and has a slightly wider contour. It is particularly striking to note how much better the results seem to be than the sensitivity in δ_{CP}, which perhaps we might have expected given the disagreement between the number of observed events in the two $1R_e$ samples and the PMNS prediction (presented in Table 6.4). We see more events than expected in the neutrino-mode $1R_e$ sample and fewer events than expected in the antineutrino-mode $1R_e$ sample, which implies more CP violation than the PMNS model allows. This causes the fit to favour $\delta_{CP}= -\pi/2$ much more strongly than any other value, and gives more significance for $\delta_{CP}= -\pi/2$ than the sensitivity tells us to expect.

It is likely that we are simply seeing a statistical fluctuation upwards in the neutrino-mode $1R_e$ sample and downwards in the antineutrino-mode $1R_e$ sample. The reason that this is important is that statistical fluctuations can go both ways. If the result we see here is due to a statistical fluctuation (and it seems that it may be),

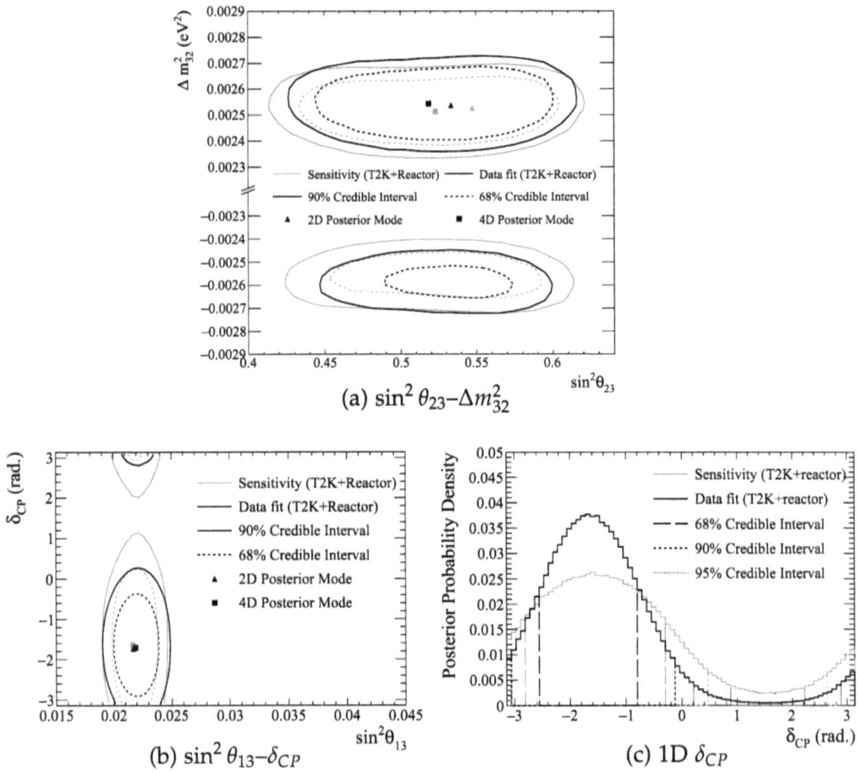

Fig. 6.19 Results of the T2K+Reactor fit compared to the T2K+Reactor Asimov fit presented in Sect. 5.5.1

then collecting more data is just as likely to make the constraint on δ_{CP} worse as it is to make it better. This highlights the need for higher statistics, so that we can be confident that the measurement is not driven by fluctuations.

In the future it would be interesting to investigate the apparent tension between the T2K data and the PMNS prediction by fitting the data in a way that does not rely on the PMNS framework. This could be achieved by fitting the oscillation probabilities directly, perhaps using a polynomial as a function of energy, or by introducing new parameters which multiply the PMNS oscillation probability (similar to the parameter β defined in the $\bar{\nu}_e$ appearance search in Appendix A.2), which would give the MCMC the freedom to move away from the PMNS prediction. The results of analyses such as these could be compared to predictions or the results of fits in the PMNS framework to give an estimate of how consistent the data are with the model. Deviations from the PMNS neutrino oscillation model could be indicative of non-standard matter effects or other new physics.

6.5 Effect of Prior

As discussed in Chap. 4, any Bayesian analysis is dependent not only on the data, but also on the prior assumptions about the model parameters that relate the likelihood to the posterior probability:

$$P(\vec{\theta}|D) = \frac{P(D|\vec{\theta})P(\vec{\theta})}{\int P(D|\vec{\theta})P(\vec{\theta})d\vec{\theta}}$$

where $P(\vec{\theta}|D)$ is the posterior probability distribution for some set of model parameters, $\vec{\theta}$, given some data, D, $P(D|\vec{\theta})$ is the likelihood, and $P(\vec{\theta})$ is the prior probability distribution for the model parameters.

It is therefore important to investigate the effect of the prior on the results. If the data constraint (i.e. the likelihood term) is very strong compared to the prior, then we expect the prior to have little effect on the posterior probability. If, however, the data constraint is weak compared to the prior, the result may change depending on which prior we choose. It is important to know if this is the case, in order to report the results in the most useful and honest way. Of course, we have already considered two choices of prior on θ_{13} in the fit: either flat in $\sin^2\theta_{13}$ (the T2K-only fit), or Gaussian in $\sin^2 2\theta_{13}$ (the T2K+Reactor fit). We have seen that the choice of prior has a significant effect on the fit results—the T2K data alone do not have strong power to constrain $\sin^2\theta_{13}$—and therefore concluded that the best approach is to show both results. It is important to consider whether we should adopt this approach for priors on the other oscillation parameters.

The MCMC method provides an easy way to investigate the effect of the prior without having to repeat the fit. Because the posterior probability is proportional to the prior probability multiplied by the likelihood, it is possible to reweight the fit chain to find the posterior distribution according to a different prior. We simply have to weight each step by the ratio of the desired prior to the prior used for the original fit. This is particularly easy in our case because the T2K-only fit uses flat priors in the interesting parameters $\sin^2\theta_{23}$, $\sin^2\theta_{13}$, Δm^2_{32}, and δ_{CP}.

The method used to reweight the chain is as follows: consider a probability distribution $f(x)$. A variable transformation to variable $y = y(x)$ gives the equivalent probability distribution $g(y)$:

$$g(y) = f(x(y))\frac{dx}{dy}$$

If $f(x)$ is a uniform distribution (i.e. a flat prior) in x then this simplifies even further:

$$g(y) \propto \frac{dx}{dy}$$

to within some constant of normalisation. In our case, y represents the 'known' variables in the posterior distribution ($\sin^2\theta_{13}$, $\sin^2\theta_{23}$, Δm^2_{32}, and δ_{CP}), and x is some variable in which we wish to have a flat prior. $g(y)$ must be evaluated as a function of the 'known' variables y. If we wish to apply a prior that is flat in $x = \theta_{23}$, instead of flat in $y = \sin^2\theta_{23}$, we must evaluate:

$$
\begin{aligned}
g(y) &\propto \frac{dx}{dy} = \left(\frac{dy}{dx}\right)^{-1} \\
&\propto \left(\frac{d}{dx}\sin^2 x\right)^{-1} \\
&\propto \frac{1}{2\sqrt{y - y^2}}
\end{aligned}
$$

This function is evaluated for each step in the MCMC chain, and its value applied as a weight to that step. This method is, of course, not perfect (you can't reweight steps you don't have), but will give us some idea of how the result depends on the prior.

Four alternative priors were considered:

1. Flat in θ_{23} (instead of flat in $\sin^2\theta_{23}$)
2. Flat in θ_{13} (instead of flat in $\sin^2\theta_{13}$)
3. Flat in $\sin^2 2\theta_{13}$ (instead of flat in $\sin^2\theta_{13}$)
4. Flat in $\sin(\delta_{CP})$ (instead of flat in δ_{CP})

The first two priors were chosen out of interest, with no particular motivation other than we perhaps have no more reason to believe that the distribution is flat in θ than we have reason to believe it is flat in $\sin^2\theta$. The flat prior in $\sin^2 2\theta_{13}$ was chosen to be more similar to the reactor neutrino experiments, which often fit in $\sin^2 2\theta_{13}$ instead of $\sin^2\theta_{13}$. Finally, the flat prior on $\sin(\delta_{CP})$ was chosen with CP violation in mind. Given that we have very little knowledge about δ_{CP}, a flat—or uninformative—prior is clearly the most suitable choice, but there is a question as to in which variable the prior should be flat. A flat prior in δ_{CP}, as we use in the fit, is appropriate and useful when one wants to make a parameter measurement (in other words, when the aim is to determine the value of δ_{CP}). However, the CP-violating terms in the oscillation probability go as $\sin(\delta_{CP})$, and we will rule out CP conservation when we can exclude $\sin(\delta_{CP}) = 0$. Therefore, if we want to talk about CP violation it may be more appropriate to consider $\sin(\delta_{CP})$, and to use a prior which is flat in that parameter.

The effect of all four priors was considered for the T2K-only fit, and the effect of priors (1) and (4) on the T2K+Reactor fit (which already includes a non-flat prior on $\sin^2 2\theta_{13}$). The results are shown in Figs. 6.20 (in $\sin^2\theta_{23}$–Δm^2_{32}) and 6.21 (in $\sin^2\theta_{13}$–δ_{CP}, not including prior (4)). It is clear from these figures that the chosen priors do not affect the results significantly. There is a very small change in the inverted-hierarchy contours in $\sin^2\theta_{23}$–Δm^2_{32} from the T2K+Reactor fit when using a flat prior in $\sin(\delta_{CP})$, and a small shift in $\sin^2\theta_{13}$ when changing from a flat prior

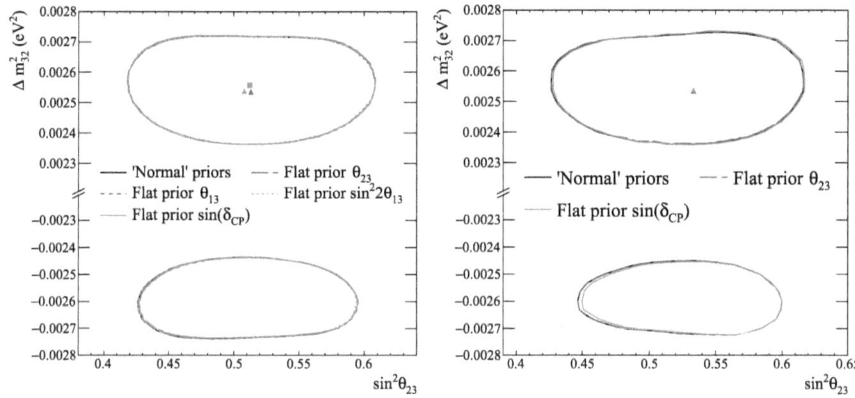

(a) T2K-only fit (b) T2K+Reactor fit

Fig. 6.20 90% credible intervals in $\sin^2\theta_{23}$–Δm^2_{32} for a number of different choices of prior: 'normal' priors (flat in $\sin^2\theta_{23}$, Δm^2_{32}, and δ_{CP}, and either flat in $\sin^2\theta_{13}$ for the T2K-only fit or Gaussian in $\sin^2 2\theta_{13}$ according to the reactor constraint for the T2K+Reactor fit), flat in θ_{23}, flat in θ_{13}, flat in $\sin^2 2\theta_{13}$, and flat in $\sin(\delta_{CP})$

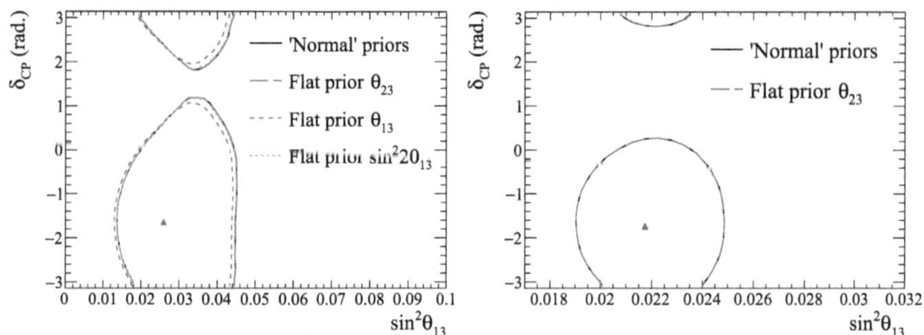

(a) T2K-only fit (b) T2K+Reactor fit

Fig. 6.21 90% credible intervals in $\sin^2\theta_{13}$–δ_{CP} for a number of different choices of prior: 'normal' priors (flat in $\sin^2\theta_{23}$, Δm^2_{32}, and δ_{CP}, and either flat in $\sin^2\theta_{13}$ for the T2K-only fit or Gaussian in $\sin^2 2\theta_{13}$ according to the reactor constraint for the T2K+Reactor fit), flat in θ_{23}, flat in θ_{13}, and flat in $\sin^2 2\theta_{13}$

in $\sin^2\theta_{13}$ to a flat prior in θ_{13}. However, these changes are not significant, and in particular do not affect our physics conclusions.

The effect of the $\sin(\delta_{CP})$ prior on the δ_{CP} distribution has been deliberately left out of Fig. 6.21 because it requires some more careful consideration. The Jacobian required to transform a prior distribution which is flat in $\sin(\delta_{CP})$ into a function of δ_{CP} is $|\cos(\delta_{CP})|$, which goes to zero and has a discontinuity in its derivative at $\delta_{CP} = \pm\pi/2$. These features of the Jacobian make the resulting posterior distribution as a function of δ_{CP} difficult to understand and interpret, and not easy to use. Therefore, it is better to present the posterior density in a way which is consistent with the

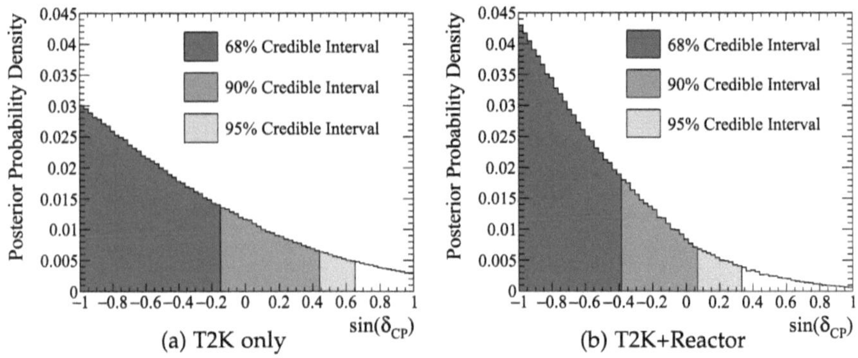

Fig. 6.22 1D posterior distribution as a function of $\sin(\delta_{CP})$ from the T2K-only and T2K+Reactor data fit, with flat prior in $\sin(\delta_{CP})$

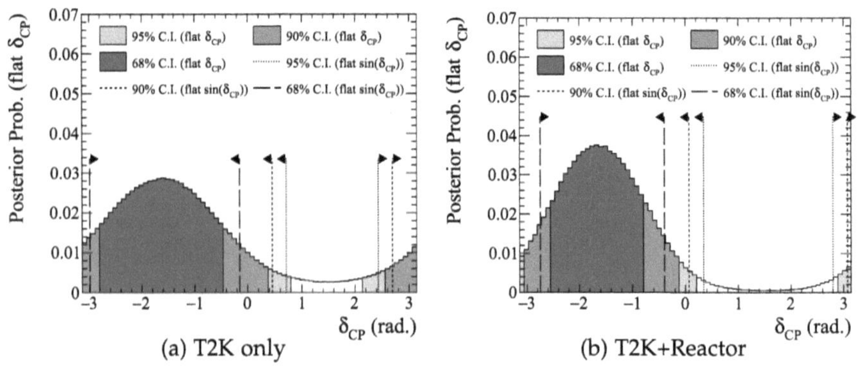

Fig. 6.23 1D posterior distribution as a function of δ_{CP} from the T2K-only and T2K+Reactor data fit, with flat prior in δ_{CP}. Credible interval limits calculated with a flat prior in $\sin(\delta_{CP})$ are overlaid

prior: as a function of δ_{CP} when a flat prior in δ_{CP} is used, and as a function of $\sin(\delta_{CP})$ when a flat prior in $\sin(\delta_{CP})$ is used. When using a flat prior in $\sin(\delta_{CP})$, the 68, 90, and 95% 1D credible interval limits on δ_{CP} can be found by taking the limits of the credible intervals in $\sin(\delta_{CP})$ and calculating both values of δ_{CP} which correspond to these limits.

Figure 6.22 shows the $\sin(\delta_{CP})$ distribution from both the T2K-only and T2K+Reactor data fits with a flat prior in $\sin(\delta_{CP})$, with credible intervals marked. Both peak at $\sin(\delta_{CP})= -1$ (corresponding to $\delta_{CP}= -\pi/2$) and go to a minimum at $\sin(\delta_{CP})= 1$ (corresponding to $\delta_{CP}= \pi/2$), and the T2K+Reactor distribution is more peaked than the T2K-only distribution. These features are consistent with what we expect from the results shown previously in this chapter.

Figure 6.23 shows the δ_{CP} distribution with a flat prior in δ_{CP} from both fits. This is the same as Fig. 6.6. The credible intervals using a flat prior in $\sin(\delta_{CP})$ are overlaid. It is clear that the prior has an effect on the final result—the credible intervals in δ_{CP} are not identical when we change from using a flat prior in δ_{CP} to a flat prior in $\sin(\delta_{CP})$. In particular, the 90% credible interval in the T2K+Reactor fit moves from excluding $\delta_{CP}=0$ and $\pm\pi$ to including those values. However, the effect is not drastic and does not significantly change our physics conclusions (except for the fairly arbitrary 90% line), so while this shows that the information provided by the T2K data set is currently not significantly stronger than the prior, neither is it significantly weaker.

References

1. K. Abe et al. (T2K Collaboration), Observation of electron neutrino appearance in a muon neutrino beam. Phys. Rev. Lett. **112**, 061802 (2014). doi:10.1103/PhysRevLett.112.061802
2. K. Abe et al. (T2K Collaboration), Measurements of neutrino oscillation in appearance and disappearance channels by the T2K experiment with 6.6×10^{20} protons on target. Phys. Rev. D **91**, 072010 (2015). doi:10.1103/PhysRevD.91.072010
3. J. Beringer et al. (Particle Data Group), Review of particle physics. Phys. Rev. D **86**, 010001 (2012) (2013 partial update for the 2014 edition), doi:10.1103/PhysRevD.86.010001
4. P. Adamson et al. (NOνA Collaboration), First measurement of muon-neutrino disappearance in NOvA. Phys. Rev. D **93**, 051104 (2016). doi:10.1103/PhysRevD.93.051104
5. R. Wendell for the Super-Kamiokande Collaboration, Atmospheric neutrino oscillations at Super-Kamiokande. Proc. Sci. (ICRC2015) **1062**, (2015), http://pos.sissa.it/archive/conferences/236/1062/ICRC2015_1062.pdf, contours taken from http://www-sk.icrr.u-tokyo.ac.jp/indico/conferenceDisplay.py?confId=2773
6. M.G. Aartsen et al. (IceCube Collaboration), Determining neutrino oscillation parameters from atmospheric muon neutrino disappearance with three years of icecube deepcore data. Phys. Rev. D **91**, 072004 (2015). doi:10.1103/PhysRevD.91.072004
7. A. Sousa (MINOS and MINOS+ Collaborations), First MINOS+ data and new results from MINOS, in *AIP Conference Proceedings* vol. 1666 (2015), p. 110004. doi:10.1063/1.4915576, *Proceedings of the Neutrino 2014 Conference*, Boston, 2–7 June 2014, contours taken from http://www-numi.fnal.gov/PublicInfo/forscientists.html

Chapter 7
Conclusions and Outlook

This thesis has presented a Bayesian statistical analysis of data from the T2K near and far detectors. This is the first T2K oscillation analysis to fit data collected with both a neutrino-mode and antineutrino-mode beam, amounting to a total of 7.00×10^{20} POT and 7.47×10^{20} POT respectively at the far detector. The use of data from the ND280 near detector significantly decreases the systematic uncertainty at the far detector, and one of the new improvements in this analysis over previous T2K oscillation analyses has been the inclusion of water-target data at ND280. This allows the cross-section systematic uncertainty for interactions on oxygen to be constrained at the near detector, which greatly reduces the uncertainty in the far detector prediction compared to previous T2K analyses which used ND280 carbon-target data only (for example, the uncertainties shown in Appendix A, Table A.1, or in [1]). The use of both $1R_\mu$ and $1R_e$ data samples from the far detector gives sensitivity to $\sin^2\theta_{23}$ and Δm^2_{32} through the disappearance channel, and sensitivity to $\sin^2\theta_{13}$ and δ_{CP} in the appearance channel. The fact that antineutrino-mode beam data are used as well as neutrino-mode data means that the analysis has sensitivity to δ_{CP} from T2K data alone, and the sensitivity is improved when combined with a prior on $\sin^2 2\theta_{13}$ from measurements by reactor neutrino experiments.

In addition to the joint $\nu + \bar{\nu}$ fit, which is the focus of this thesis, two analyses using only antineutrino-mode data at Super-Kamiokande have been completed: a measurement of the antineutrino oscillation parameters $\sin^2\bar{\theta}_{23}$ and $\Delta\bar{m}^2_{32}$ from $\bar{\nu}_\mu$ disappearance—the first at T2K—and the first ever search for $\bar{\nu}_e$ appearance. There was not space to discuss these analyses in the main body of this thesis, but the methods and results are summarised in Appendix A.

The results of the joint $\nu + \bar{\nu}$ oscillation fit are summarised in Table 7.1, where the uncertainties given represent the 1D 68% credible interval limits. For δ_{CP}, the 90% 1D credible interval using a flat prior in δ_{CP} from the T2K-only fit finds $\delta_{CP} \notin [0.38, 2.60]$, but includes the CP-conserving values $\delta_{CP} = 0$ and $\pm\pi$. The 90% credible interval using a flat prior in δ_{CP} from the T2K+Reactor fit excludes the

K.E. Duffy, *First Measurement of Neutrino and Antineutrino Oscillation at T2K*,
Springer Theses, DOI 10.1007/978-3-319-65040-1_7

Table 7.1 Best-fit values and 68% credible interval uncertainties from the T2K-only and T2K+Reactor ν+$\bar{\nu}$ data fits. No uncertainties are given for δ_{CP} because the distribution is non-Gaussian and cyclical, but the 90% 1D credible interval limits are given in the text

	Δm^2_{32} ($\times 10^{-3}$eV^2)	$\sin^2\theta_{23}$	$\sin^2\theta_{13}$	δ_{CP}
T2K-only	$2.54^{+0.12}_{-0.10}$	$0.519^{+0.031}_{-0.059}$	$0.0257^{+0.0106}_{-0.0440}$	-1.91
T2K+Reactor	$2.54^{+0.12}_{-0.12}$	$0.519^{+0.061}_{-0.029}$	$0.0219^{+0.0014}_{-0.0014}$	-1.72

CP-conserving values, and includes the range $\delta_{CP} \in [-3.10, -0.17]$. This is, however, somewhat prior-dependent—the CP-conserving values of δ_{CP} are not excluded at 90% in the T2K+Reactor fit when a flat prior in $\sin(\delta_{CP})$ is used rather than a flat prior in δ_{CP}. The results in terms of the other oscillation parameters $\sin^2\theta_{23}$, Δm^2_{32}, and $\sin^2\theta_{13}$ do not seem to depend strongly on the chosen prior, with the exception of the choice of whether or not to use the reactor measurement as a prior on $\sin^2 2\theta_{13}$. Both the T2K-only and T2K+Reactor fits show a preference for the normal mass hierarchy and the upper octant, but the preference is not statistically significant.

The results presented here are the most precise measurements of $\sin^2\theta_{23}$, Δm^2_{32}, $\sin^2\theta_{13}$, and δ_{CP} at T2K. They are consistent with other oscillation analyses on T2K, with previous T2K analyses, and with measurements by other neutrino oscillation experiments. The constraint on δ_{CP} is world-leading—no other experiment has ever seen 90% evidence of CP violation in the neutrino sector—but it is important to bear in mind the dependence of this constraint on the prior. T2K will continue running to collect data towards a final goal of 7.8×10^{21} POT (slightly over 5 times the current data set), with an increasing rate of POT collection thanks to increased beam power. The predicted sensitivity of T2K with the full data set of 7.8×10^{21} POT is discussed in [2]: the continued analysis of ν and $\bar{\nu}$ oscillation at higher statistics will provide ever more precise measurements of the oscillation parameters $\sin^2\theta_{23}$, Δm^2_{32}, $\sin^2\theta_{13}$, and δ_{CP}. Although we expect the precision to improve in the long term, if the results shown here are due to a 'lucky' statistical fluctuation it is possible that we will see the constraint on δ_{CP} get 'worse' in the short term as more statistics are collected.

There is a lot of scope to improve this analysis in the future. New data samples are currently being prepared which will reduce both statistical and systematic uncertainties: replica-target data from NA61/SHINE (in which measurements were made using a replica of the T2K target) will soon be included to reduce uncertainty due to the neutrino flux prediction; an improved selection is being implemented for ND280 which will accept backwards-going and high-angle muon tracks (the current selection accepts forwards-going tracks only); and a selection for ν_e interactions with an electron and a π^+ in the final state has been developed at SK, which will improve the ν_e statistics at the far detector.

Further structural changes in the analysis could also lead to improvements. The SK data samples are currently binned in reconstructed neutrino energy, but more information could be included, for example by binning in two variables: either by adding the momentum, p, or angle, θ, of the outgoing lepton to the reconstructed energy, or by binning in p and θ as is done for the ND280 data samples. The additional degree of freedom should allow for a more accurate signal/background distinction and a more precise measurement of the oscillation parameters.

The uncertainty due to final state and secondary interactions at SK is now one of the dominant uncertainties in the analysis (as shown in Tables 5.7–5.9), so the treatment of these systematic parameters will require careful consideration in the future.

In addition to these improvements to the T2K analysis, vast gains in the determination of oscillation parameters may be made by combining T2K data with that of other neutrino experiments. To this end, a combined fit with the NOνA experiment is under preliminary discussion. Combining the data sets of the two experiments will allow stronger limits to be set on δ_{CP} and the neutrino mass hierarchy (to which NOνA has more sensitivity than T2K, due to its longer baseline).

Aside from improving the analysis framework, future work in this area may include fitting the T2K data outside the PMNS framework, in order to test the PMNS model. It seems that the data show fluctuations away from the PMNS prediction—even with maximal CP violation—in both the neutrino-mode and antineutrino-mode $1R_e$ samples. Fitting the oscillation probabilities directly could give an insight into how compatible the data are with the theory as well as providing an alternative way to highlight CP violation in the neutrino sector.

References

1. K. Abe et al. (T2K Collaboration), Measurements of neutrino oscillation in appearance and disappearance channels by the T2K experiment with 6.6×10^{20} protons on target. Phys. Rev. D **91**, 072010 (2015). doi:10.1103/PhysRevD.91.072010
2. K. Abe et al. (T2K Collaboration), Neutrino oscillation physics potential of the T2K experiment. Prog. Theor. Exp. Phys. **2015**(4), (2015). doi:10.1093/ptep/ptv031

Appendix A
Antineutrino Oscillation: Search for $\bar{\nu}_e$ Appearance and Measurement of $\bar{\nu}_\mu$ Disappearance

The main focus of this thesis is an oscillation analysis in which both neutrino and antineutrino T2K data are fit, using a Markov Chain Monte Carlo method, to find estimates of the oscillation parameters $\sin^2\theta_{23}$, Δm^2_{32}, $\sin^2\theta_{13}$, and δ_{CP}. The analysis assumptions, framework, and validations are described in Chap. 5, and the results of fitting the T2K data are shown in Chap. 6. However, two further oscillation analyses have also been completed using only the antineutrino-mode data: a search for $\bar{\nu}_e$ appearance and a measurement of $\bar{\nu}_\mu$ disappearance at T2K. Unfortunately, space does not permit an in-depth description of all three analyses, so instead this appendix hopes to summarise the considerations and results for the two antineutrino analyses, in order to allow for a full discussion of the joint neutrino-antineutrino fit in Chaps. 5 and 6. A large amount of this summary has been adapted from [1], and further details about the analyses (including validations, discussions of the models used, and further discussions of the results) can be found in [2, 3]. The $\bar{\nu}_\mu$ disappearance measurement has also been published in Physical Review Letters [4].

There are three main differences between the antineutrino analyses discussed in this appendix and the joint neutrino-antineutrino fit framework discussed in Chap. 5. Firstly, the far detector data set is different: the $\bar{\nu}_e$ appearance and $\bar{\nu}_\mu$ disappearance analyses use only the data from one year of T2K antineutrino-mode running, corresponding to 4.01×10^{20} POT, and only one SK data selection for each analysis—either the single-ring e-like (RHC 1R_e) sample (for the $\bar{\nu}_e$ appearance search), or the single-ring μ-like (RHC R_μ) sample (for the $\bar{\nu}_\mu$ disappearance measurement). Secondly, the near detector data set is different. These analayses use less antineutrino-mode data than the analysis described in Chaps. 5 and 6 (the ND280 data set used here amounts to 5.82×10^{20} POT in neutrino mode and 0.43×10^{20} POT in antineutrino mode), but the largest difference in the ND280 samples is that only data for interactions in FGD1 are included. This means that there are no interactions on oxygen at ND280, so the ND280 data are unable to constrain oxygen-specific cross-section parameters. This results in larger cross-section uncertainties at SK, which is why a large amount of work was done to include FGD2 data in the neutrino-antineutrino joint fit. For both SK and ND280 the samples and selections are the same as those described in Sect. 5.2, but are repeated briefly below.

K.E. Duffy, *First Measurement of Neutrino and Antineutrino Oscillation at T2K*, Springer Theses, DOI 10.1007/978-3-319-65040-1

The final difference between the antineutrino analyses and the analysis described in Chaps. 5 and 6 is the cross-section model—a number of small updates were made to the cross-section model used in this appendix before the neutrino-antineutrino joint fit. The cross-section model used for the antineutrino analyses is described in [5,6], and the updates for the neutrino-antineutrino joint fit are described in [7, 8].

A.1 Analysis Strategy

The analysis strategy for both results presented here is similar to previous T2K results [9] and the neutrino/antineutrino joint fit described in Chaps. 5 and 6: data samples of charged current (CC) interactions are selected at ND280 and SK and fit simultaneously to make a measurement of neutrino oscillation. The analysis takes inputs from theoretical flux, cross-section, and detector models, each of which has its own uncertainties, as well as assuming the PMNS three-flavour neutrino oscillation model. Both analyses described in this appendix use the same ND280 data and models, but different assumptions and data samples at the far detector.

The predictions at both ND280 and SK use the same flux and cross-section models, so the ND280 data can reduce the uncertainty in the SK prediction by constraining parameter values in the underlying models. Information from external data (from the MINERνA, MiniBooNE, and ANL and BNL bubble chamber experiments) are also included as priors on the cross-section parameters [5, 6]. When using only FGD1 data from ND280, it is not possible for the near detector to constrain all cross-section parameters because of the different target materials in the near detector (primarily carbon) and far detector (primarily oxygen), as the relative uncertainties between interactions on carbon and oxygen are not always well understood. Because of this, separate parameters are used to describe Fermi momentum, binding energy, multinucleon event normalisation and CC coherent pion production normalisation on oxygen, which are not well constrained by the near detector. A conservative (100% uncertainty, with no correlation with multinucleon events on carbon) Ansatz is used for the multinucleon (2p-2h, or MEC) normalisation on oxygen.

The data in the near detector are fit in the momentum and angle of the outgoing lepton from the neutrino interaction. Data from the neutrino-mode beam as well as antineutrino mode are used to ensure that the model parameters are consistent between neutrinos and antineutrinos, and provide a constraint on the wrong-sign background (ν in the $\overline{\nu}$ beam). The antineutrino-mode data are split into 'ν_μ CC 1-track', 'ν_μ CC N-track', '$\overline{\nu}_\mu$ CC 1-track', and '$\overline{\nu}_\mu$ CC N-track' samples. The '1-track' samples are dominated by CC quasielastic (CCQE) interactions ($\nu_\mu + n \rightarrow \mu^- + p$ or $\overline{\nu}_\mu + p \rightarrow \mu^+ + n$, the 'signal' at Super-Kamiokande), and the 'N-track' samples are used to estimate the background from other interactions. The neutrino-mode data are divided into three subsamples according to the number of measured pions associated with the interaction: 'ν_μ CC0π', 'ν_μ CC1π', and 'ν_μ CC other', which are dominated by CCQE, CC resonant pion production, and CC deep inelastic scattering interactions respectively.

The data from Super-Kamiokande are fit in reconstructed antineutrino energy, as defined in Eq. 3.1 in Chap. 3, assuming that the interaction was a CCQE interaction of a neutrino or antineutrino with a bound nucleon at rest. In both the $1R_e$ and R_μ event selections, events must be fully contained within the fiducial volume of the detector, and have only one reconstructed ring (hence "single-ring" in the selection name). For the $1R_e$ selection, we further require that the outgoing lepton from the interaction has particle identification consistent with an electron and energy greater than that of a 100 MeV electromagnetic shower. We also require that the reconstructed antineutrino energy is smaller than 1.25 GeV, that the event passes the π^0 rejection cuts used in the Super-Kamiokande software, and that there are no decay electrons. The event selection for the R_μ sample requires that the event is consistent with a muon-like particle identification, there must be one or fewer decay electrons, and the outgoing lepton momentum must be greater than 200 MeV.

Because the data set at ND280 is much larger than at SK, the ND280 data dominate the fit for the flux and cross-section systematic parameters. Therefore the behaviour of these parameters in the fit is similar for both analyses described in this appendix, because they use the same ND280 data even though they use different SK data samples. Figure A.1 shows some of the flux and cross-section parameters with their associated uncertainties before and after the $\bar{\nu}_\mu$ disappearance fit described in Sect. A.3. The predicted flux at Super-Kamiokande is generally increased by the fit, and the uncertainty is decreased. Some of the cross-section parameters fit to values which are significantly different to their prior predictions, particularly the multinucleon event normalisation parameter on carbon ("MEC (C)"). This was part of the motivation for updating the cross-section model for the joint neutrino-antineutrino fit discussed in Chaps. 5 and 6. The uncertainties on the parameters to which the near detector is sensitive are generally decreased in the fit, but there is not as dramatic a reduction in the uncertainties of the oxygen-specific parameters. This is because of the low statistics in the Super-Kamiokande data samples (the R_μ sample used for this analysis contains only 34 events) and the fact that the oxygen-specific parameters are not constrained at ND280.

Since we see that the largest reduction in systematic uncertainty comes from the ND280 data, rather than the SK data, it is illuminating to consider the behaviour of these systematic uncertainties before and after a fit to the ND280 data only. Table A.1 shows the uncertainty in the predicted number of events in the Super-Kamiokande RHC R_μ sample due to each source of systematic uncertainty, before and after a fit to only the ND280 data. The near detector data reduces the uncertainty due to the parameters that it can constrain from 9.2 to 3.4%. However, the overall uncertainty is dominated by the cross-section parameters which are not constrained by the near detector (in particular the multinucleon event normalisation parameter on oxygen, which alone produces a 9.5% uncertainty), and is only reduced from 14.4 to 11.6% by the ND280 data.

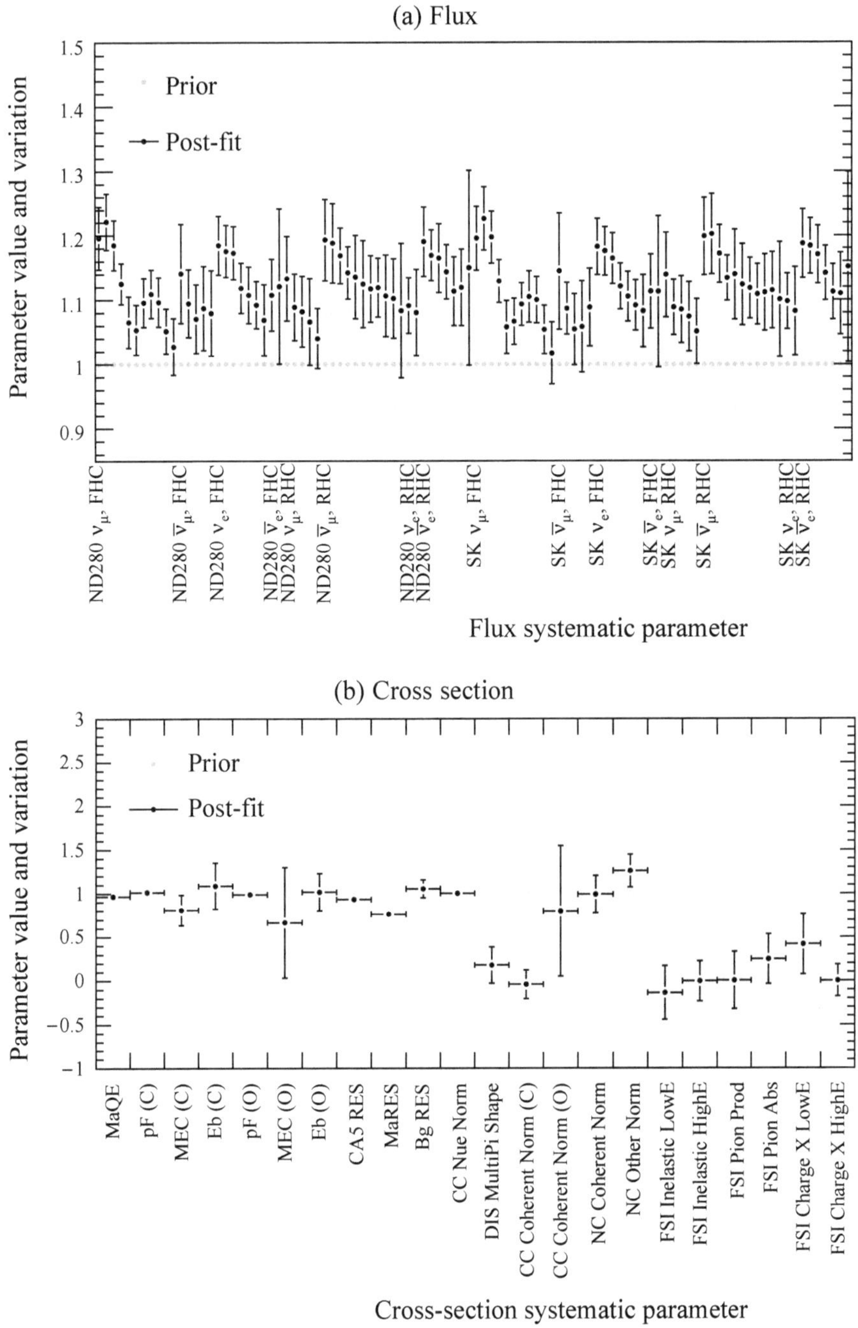

Fig. A.1 Central values and 1σ uncertainties for the flux and cross-section systematic parameters before and after the $\bar{\nu}_\mu$ disappearance fit

Table A.1 Fractional uncertainty in the number of events predicted in the SK RHC R_μ sample due to different sources of systematic uncertainty, before and after a fit to the near detector (ND) data

Systematic	$\Delta N_{SK}/N_{SK}$ before ND fit	$\Delta N_{SK}/N_{SK}$ after ND fit
All common to ND/SK	9.2%	3.4%
Multinucleon effect on oxygen	9.5%	
All oxygen cross-section	10.0%	
FSI/SI at SK	2.1%	
SK detector	3.8%	
Total	14.4%	11.6%

A.2 $\bar{\nu}_e$ Appearance Analysis

This section presents the first ever search for $\bar{\nu}_e$ appearance on T2K, in which the aim is to find evidence for electron antineutrino appearance, separately from electron neutrino appearance (which has already been observed at T2K with a significance of 7.3σ [10]). Because of the large wrong-sign contamination in the RHC neutrino beam, it is important to correctly account for the fact that there will be a ν_e appearance background to the desired $\bar{\nu}_e$ appearance signal in the SK RHC $1R_e$ sample. Therefore, in order to test whether or not the data indicate the presence of $\overline{\nu}_e$ appearance, a new parameter β is introduced, which modifies the $\overline{\nu}_e$ appearance probability:

$$P(\overline{\nu}_\mu \rightarrow \overline{\nu}_e) = \beta \times P_{PMNS}(\overline{\nu}_\mu \rightarrow \overline{\nu}_e)$$

β is defined to have value either 0 or 1, such that $\beta = 1$ corresponds to $\overline{\nu}_e$ appearance in accordance with the PMNS prediction (which allows for CP violation if $\delta_{CP} \neq 0$), and $\beta = 0$ corresponds to no $\overline{\nu}_e$ appearance. Aside from this, CPT symmetry is assumed (the same oscillation parameters are used for neutrino and antineutrino oscillations).

The analysis uses a marginal likelihood, which is integrated over all parameters other than β:

$$\mathcal{L}(\beta) = \int\int \prod_{bins} \mathcal{L}_{Poisson,bin}(D|\beta, \vec{o}, \vec{f}) \times P_{syst.}(\vec{f}) \times P_{osc.}(\vec{o}) d\vec{o} d\vec{f}$$

where $\vec{o}$ are the oscillation parameters, $\vec{f}$ are the systematic parameters, $P_{syst.}(\vec{f})$ is the prior probability density for the systematic parameters, and $P_{osc.}(\vec{o})$ is the prior probability density for the oscillation parameters. $\mathcal{L}_{Poisson,bin}$ is the Poisson likelihood in each bin given the number of data events, D, and the number of predicted events at ND280 and Super-Kamiokande, determined by $\vec{o}$ and $\vec{f}$. The product runs over all ND280 and SK bins. The oscillation priors are taken from the posterior of the T2K joint ν_μ and ν_e fit [9], which have a peak value for $\delta_{CP} \sim -\pi/2$.

The result of this analysis is a Bayes factor for $\beta = 1$ over $\beta = 0$, B_{10}. This is simply the likelihood ratio:

$$B_{10} = \frac{\mathcal{L}(D|\beta = 1)}{\mathcal{L}(D|\beta = 0)}$$

The Bayes factor describes how much the data favour $\beta = 1$ over $\beta = 0$, and is the factor by which we must 'update our beliefs' based on the data, D (i.e. the factor by which we must multiply the prior odds of two hypotheses to find the posterior odds). For example, imagine that the fit result gives Bayes factor $B_{10} = 2$. If we gave both $\beta = 0$ and $\beta = 1$ hypotheses equal a priori odds, we would find a posterior odds ratio of $2 \times 1 = 2$. This is equivalent to a posterior probability of 66.7% for $\beta = 1$, compared to a prior probability of 50% (since the only possibilities for β are 0 and 1 in this model, we require the posterior probabilities $P(\beta = 0|D) + P(\beta = 1|D) = 1$). However, one might argue that this is a strange choice of prior odds for our situation: very few physicists would agree that it's equally likely for $\bar{\nu}_e$ appearance to occur or not occur. Therefore consider a different prior, one in which the prior odds ratio for $(\beta = 1)/(\beta = 0)$ is 10 (meaning we believe *a priori* that $\beta = 1$ is 10 times more likely to be true than $\beta = 0$). In this case a Bayes factor of 2 would give a posterior ratio of $2 \times 10 = 20$, or a posterior probability of 95.2% (compared to a prior probability of 90.9%).

The data set used for this analysis contains 3 events in the 1R$_e$ sample at Super-Kamiokande. At the oscillation parameter values given in Table A.2, we expect ~1.3 events if $\beta = 0$ and ~3.7 events if $\beta = 1$. Examining the full range $\delta_{CP} = [-\pi, \pi]$ and both mass hierarchies, we predict between 3.7 and 5.3 events for $\beta = 1$.

The Bayes factor for this data fit is 1.1, which does not show strong evidence in favour of the $\beta = 1$ hypothesis over $\beta = 0$. This may perhaps be understood by examining the posterior distribution as a function of reconstructed energy. Figure A.2 show this distribution, found using the posterior predictive method described in Sect. 4.4.5. The posterior predictive procedure was repeated three times: once sampling randomly from the entire posterior distribution (essentially marginalising over β), once sampling only from steps in the posterior distribution with $\beta = 1$, and once sampling only from steps with $\beta = 0$. This enables us to examine the overall posterior distribution as well as the best-fit spectra for the two models individually. The result of sampling randomly from the entire posterior distribution is shown as the 2D histogram in Fig. A.2, and the best-fit reconstructed energy spectra assuming $\beta = 0$ and $\beta = 1$ are overlaid using a dashed and solid line respectively. This gives some insight into why the analysis is not able to choose a preferred model: with this amount of data it is impossible to resolve the posterior probability distributions as a function of reconstructed energy for the $\beta = 0$ and $\beta = 1$ models, which means that for many reconstructed energies both hypotheses may be equally consistent with the data. With more antineutrino-mode data we would expect the gap between the solid and dashed lines to grow, giving the fit more power to choose between the models.

The Bayes factor presented here is consistent with the results found by the two other analysis groups on T2K, which use a hybrid frequentist-Bayesian framework. One of those analyses, which also bins the SK data and prediction in reconstructed

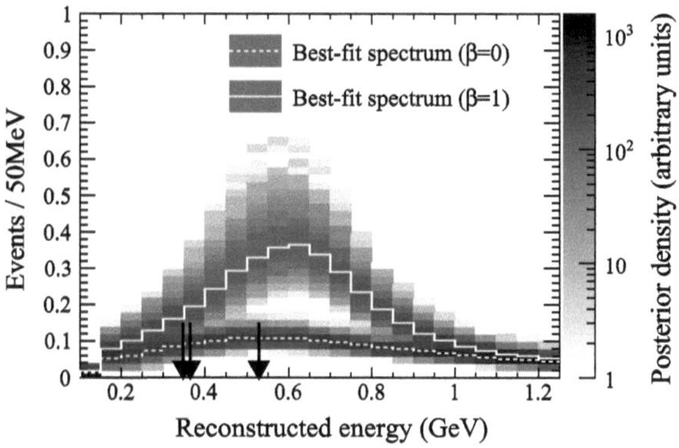

Fig. A.2 Best-fit reconstructed energy spectrum for the SK RHC 1R$_e$ sample in the $\bar{\nu}_e$ appearance analysis. The result of sampling the fit posterior 2500 times, marginalising over β, is shown as the 2D histogram and the best-fit spectra for the $\beta = 0$ and $\beta = 1$ models are overlaid. The reconstructed energies of the three observed events are also indicated by the *black arrows*

antineutrino energy, found a frequentist p-value of 0.16 and a marginal likelihood ratio (equivalent to the Bayes factor) of 1.1. The other analysis, which bins the SK data and prediction in the momentum and angle of the charged lepton produced in the neutrino interaction, found a frequentist p-value of 0.34 and a marginal likelihood ratio of 0.6 (which is equivalent to a Bayes factor of $\sim$1.7 in favour of $\beta = 0$ over $\beta = 1$). All of these results are consistent: the p-values indicate no disagreement between the data and the $\beta = 0$ hypothesis, and the Bayes factors are all sufficiently close to 1 that they do not provide significant evidence in favour of either hypothesis. With this data set, we cannot conclude that we have observed $\bar{\nu}_e$ appearance with statistical significance.

A.3 $\bar{\nu}_\mu$ Disappearance Measurement

The second analysis presented in this appendix is a measurement of $\bar{\nu}_\mu$ disappearance. It uses the antineutrino-mode T2K data to measure the antineutrino oscillation parameters, which are not assumed to be equal to the neutrino oscillation parameters (i.e. CPT invariance is not assumed). This essentially doubles the number of oscillation parameters considered: we assume that neutrino oscillation is governed by the parameters $\sin^2\theta_{12}$, Δm^2_{21}, $\sin^2\theta_{23}$, Δm^2_{32}, $\sin^2\theta_{13}$, and δ_{CP}, but that the oscillation of antineutrinos is determined by the entirely independent parameters $\sin^2\bar{\theta}_{12}$, $\Delta\bar{m}^2_{21}$, $\sin^2\bar{\theta}_{23}$, $\Delta\bar{m}^2_{32}$, $\sin^2\bar{\theta}_{13}$, and $\bar{\delta}_{CP}$. If the antineutrino oscillation parameters are found to be different from their neutrino counterparts, that would suggest $P(\nu_\mu\rightarrow\nu_\mu)\neq P(\bar{\nu}_\mu\rightarrow\bar{\nu}_\mu)$, which could be indicative of CPT violation or non-standard matter effects.

Table A.2 Oscillation parameters used for the $\bar{\nu}_\mu$ disappearance analysis. The fixed values of $\sin^2\theta_{23}$, Δm^2_{32}, $\sin^2\theta_{13}$, and δ_{CP} are taken from previous T2K results [9]. The values of $\sin^2\theta_{12}$ and Δm^2_{32} are taken from the 2014 edition of the Particle Data Booklet [11]

Oscillation parameter	ν	$\bar{\nu}$
$\sin^2\theta_{23}$	0.527	0–1
$\Delta m^2_{32}(\times 10^{-3}\text{eV}^2)$	2.51	0–20
$\sin^2\theta_{13}$	0.0248	
δ_{CP} (rad.)	−1.55	
$\sin^2\theta_{12}$	0.304	
$\Delta m^2_{21}(\times 10^{-5}\text{eV}^2)$	7.53	

At the T2K baseline and neutrino energies, two parameters dominate the muon antineutrino survival probability: $\sin^2\bar{\theta}_{23}$ and $\Delta\bar{m}^2_{32}$. These two parameters are allowed to vary with a flat prior in the ranges given in Table A.2, and the other antineutrino oscillation parameters, $\sin^2\bar{\theta}_{13}$, $\sin^2\bar{\theta}_{12}$, $\Delta\bar{m}^2_{12}$, and $\bar{\delta}_{CP}$, are fixed to be identical to their matter counterparts. These and all neutrino oscillation parameters are fixed to the best-fit values from previous T2K fits [9] or the values given in the 2014 PDG [11], which ensures that the predicted neutrino background at the far detector is consistent with current world knowledge, while we use the antineutrino-mode data to estimate $\sin^2\bar{\theta}_{23}$ and $\Delta\bar{m}^2_{32}$. The values given in Table A.2 may differ from the values quoted from [9] in Chap. 5; this is because the results of two analyses were reported in [9], with slightly different—but consistent—best-fit values for the oscillation parameters. Fixing the neutrino oscillation parameters is suitable in this case because the analysis is entirely statistics-dominated. However, in future analyses with larger data sets it would be more appropriate to vary the neutrino oscillation parameters within the uncertainties given in the PDG, or (as has now been implemented at T2K in an update to this analysis) include all four SK data samples and allow the FHC data to constrain the neutrino oscillation parameters.

There are 34 events in the R_μ sample from antineutrino beam data used for this analysis. Figure A.3 shows the predicted reconstructed energy spectrum under the no-oscillation hypothesis, and the best-fit spectrum from the data fit, along with the data. The lower plot shows the ratio of both data and best-fit spectrum to the unoscillated prediction, which shows the characteristic 'oscillation dip' that is clear evidence of $\overline{\nu}_\mu$ oscillation. The binning shown in the plot is coarser than the binning used for the data fit to reduce the effect of statistical fluctuations in the figure.

The left-hand side of Fig. A.4 shows the 68 and 90% credible interval contours in $\sin^2\bar{\theta}_{23}$–$\Delta\bar{m}^2_{32}$ compared to the credible interval contours from the T2K ν_μ+ν_e oscillation fit [9]. The antineutrino analysis has much larger contours because it has much lower statistics than the neutrino analysis, but the results are consistent: we see no evidence for CPT violation or non-standard matter effects. Projecting the posterior probability distribution onto one dimension gives 68% credible interval uncertainties on the best-fit estimates:

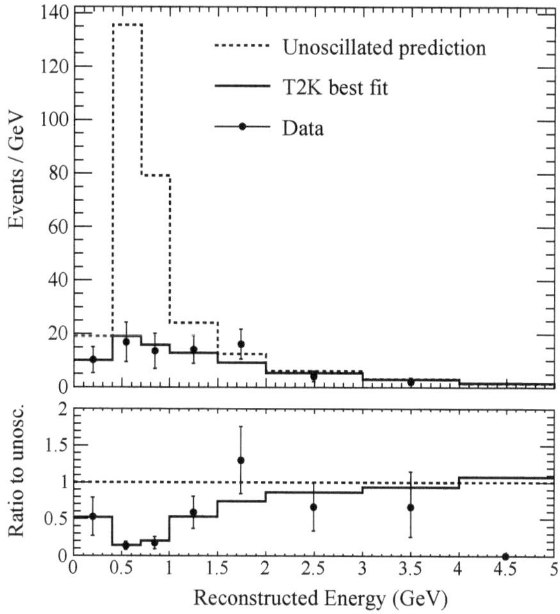

Fig. A.3 *Top* Best-fit reconstructed energy spectrum for the SK RHC R_μ sample in the $\bar{\nu}_\mu$ disappearance analysis, compared to the prediction in the case of no neutrino oscillation. Data points are overlaid in black, with statistical uncertainty. *Bottom* Ratio of best-fit energy spectrum and data to the prediction without oscillations

$$\sin^2\overline{\theta}_{23} = 0.46^{+0.14}_{-0.06}$$

$$\Delta\overline{m}^2_{32} = 2.50^{+0.3}_{-0.2} \times 10^{-3}\ \mathrm{eV}^2$$

The right-hand side of Fig. A.4 shows the credible interval contours from this analysis overlaid with the 90% confidence level contours from $\bar{\nu}_\mu$ disappearance analyses by the MINOS [12] and Super-Kamiokande collaborations [13] which were shown in Fig. 2.13 of Chap. 2. The results of all three measurements are consistent, although—as in Sect. 6.3—it is important to bear in mind that credible intervals and confidence intervals are not directly comparable. MINOS and T2K favour non-maximal best-fit points of $\sin^2\bar{\theta}_{23}$, and have a tighter constraint on $\Delta\bar{m}^2_{32}$ than Super-Kamiokande, and Super-Kamiokande and T2K have tighter constraints on $\sin^2\bar{\theta}_{23}$ than MINOS, but all three results are in agreement.

To conclude, this appendix has briefly presented two antineutrino oscillation analyses which were performed on a subset of the T2K antineutrino data, as part of the preparation for (and validation of) the joint neutrino-antineutrino analysis presented in Chaps. 5 and 6. Both analyses are statistics-limited and will be improved with additional antineutrino data in the future.

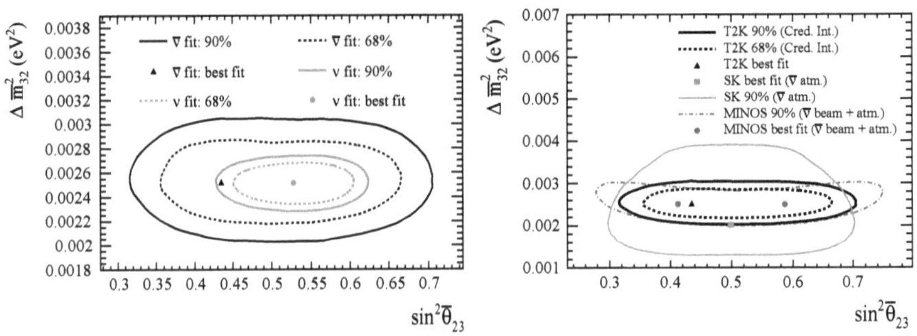

Fig. A.4 68 and 90% credible intervals in $\sin^2\bar{\theta}_{23}$–$\Delta\bar{m}^2_{32}$ from the $\bar{\nu}_\mu$ disappearance analysis. *Left* overlaid with credible interval contours from the T2K $\nu_\mu + \nu_e$ joint fit [9]. *Right* overlaid with 90% confidence level contours from similar analyses by the MINOS (using antineutrino-mode beam and atmospheric data) [12] and Super-Kamiokande collaborations (using atmospheric data only) [13]. The MINOS contour was originally presented in terms of $\sin^2 2\bar{\theta}_{23}$ and had to be transformed to $\sin^2\theta_{13}$ (hence the two best-fit points)

References

1. K.E. Duffy for the T2K Collaboration, Results and prospects from T2K, in *Proceedings from XVII International Workshop on Neutrino Factories and Future Neutrino Facilities (NuFact15), Rio de Janeiro*, vol. eConf C1508102 (2016), http://www.slac.stanford.edu/econf/C1508102/separado/2.pdf
2. A. Kaboth, K.E. Duffy, R. Calland, D. Payne, Analysis of $\bar{\nu}_e$ appearance using MCMC, Technical Report 247, T2K (2015), http://www.t2k.org/docs/technotes/247
3. A. Kaboth, K.E. Duffy, R. Calland, D. Payne, Analysis of $\bar{\nu}_\mu$ disappearance using MCMC, Technical Report 241, T2K (2015), http://www.t2k.org/docs/technotes/241
4. K. Abe et al. (T2K Collaboration), Measurement of muon antineutrino oscillations with an accelerator-produced off-axis beam. Phys. Rev. Lett. **116**, 181801 (2016). doi:10.1103/PhysRevLett.116.181801
5. A. Bercellie et al., Cross section parameters for 2014 oscillation analysis, Technical Report 192, T2K (2014), http://www.t2k.org/docs/technotes/192
6. A. Bercellie et al., Implementation of additional NIWG cross section parameterizations for 2014 analyses, Technical Report 193, T2K (2014), http://www.t2k.org/docs/technotes/193
7. M. Dunkman et al., Updated recommendation of the 2014-5 oscillation parameters, Technical Report 265, T2K (2016), http://www.t2k.org/docs/technotes/265
8. K.E. Duffy, A. Kaboth, A joint ND280-SK 1R$_\mu$-SK 1R$_e$ fit of neutrino and antineutrino-mode data using MCMC, Tech Report 269, T2K (2016), http://www.t2k.org/docs/technotes/269

9. K. Abe et al. (T2K Collaboration), Measurements of neutrino oscillation in appearance and disappearance channels by the T2K experiment With 6.6×10^{20} protons on target. Phys. Rev. D **91**, 072010 (2015). doi:10.1103/PhysRevD.91.072010
10. K. Abe et al. (T2K Collaboration), Observation of electron neutrino appearance in a muon neutrino beam. Phys. Rev. Lett. **112**, 061802 (2014). doi:10.1103/PhysRevLett.112.061802
11. K.A. Olive et al. (Particle Data Group), Review of particle physics. Chin. Phys. **C38**, 090001 (2014). (and 2015 update), doi:10.1088/1674-1137/38/9/090001
12. P. Adamson et al. (MINOS Collaboration), Combined analysis of ν_μ disappearance and $\nu_\mu \rightarrow \nu_e$ appearance in MINOS using accelerator and atmospheric neutrinos. Phys. Rev. Lett. **112**, 191801 (2014). doi:10.1103/PhysRevLett.112.191801
13. K. Abe et al. (Super-Kamiokande Collaboration), Search for differences in oscillation parameters for atmospheric neutrinos and antineutrinos at super-kamiokande. Phys. Rev. Lett. **107**, 241801 (2011). doi:10.1103/PhysRevLett.107.241801

Appendix B
Oscillation Analysis Groups on T2K

There are three oscillation analysis groups on T2K, named MaCh3 (described in this thesis), VaLOR, and P-Theta. All three groups share the same inputs (Monte Carlo predictions, parameterisation of systematic uncertainties, and pre-fit uncertainties applied to systematic parameters) but have independent fitting frameworks. Therefore, comparisons between the three groups are a very valuable tool to validate the implementation of systematic uncertainties and MC predictions, as well as the fitters themselves.

The main differences between the fitting frameworks used by the three groups are summarised below.

Binning used for SK samples

- *MaCh3*: 1D E_{rec}
- *VaLOR*: 2D E_{rec}–θ
- *P-Theta*: 2D p–θ

Treatment of near detector

- *MaCh3*: Simultaneous fit with far detector
- *VaLOR and P-Theta*: ND-only fit from the BANFF group propagated to SK and used as an input to an SK-only fit

Treatment of systematic parameters

- *All analyses*: Marginalisation

Fitting method

- *MaCh3*: MCMC
- *VaLOR and P-Theta*: Minimisation of parameters of interest, with nuisance parameters marginalised using Monte Carlo sampling

K.E. Duffy, *First Measurement of Neutrino and Antineutrino Oscillation at T2K*,
Springer Theses, DOI 10.1007/978-3-319-65040-1

Statistical approach

- *MaCh3*: Bayesian
- *VaLOR and P-Theta*: Hybrid Frequentist-Bayesian

Primary result

- *MaCh3*: Credible intervals
- *VaLOR and P-Theta*: Confidence intervals

Curriculum Vitae

Kirsty E.Duffy

RESEARCH INTERESTS	• Neutrino physics, neutrino oscillation, CP violation in the neutrino sector • Bayesian statistical methods for data analysis in high-energy physics, the Markov Chain Monte Carlo technique • DAQ systems and hardware for particle detectors
EDUCATION	**University of Oxford**, Oxford, UK PhD in Particle Physics; October 2016 Thesis title: *Measurement of the neutrino oscillation parameters* $\sin^2\theta_{23}$, Δm^2_{32}, $\sin^2\theta_{13}$, and δ_{CP} *in neutrino and antineutrino oscillation at T2K* **University of Oxford**, Oxford, UK MPhys in Physics, 1st Class; June 2012
ACADEMIC AND WORK EXPERIENCE	**University of Oxford**, Oxford, UK *Postdoctoral Research Assistant* **October 2016–June 2017** Research on the T2K experiment, expanding on the

K.E. Duffy, *First Measurement of Neutrino and Antineutrino Oscillation at T2K*,
Springer Theses, DOI 10.1007/978-3-319-65040-1

methods developed in my doctoral research to include a direct comparison of the oscillation probabilities inferred from the T2K data to the PMNS prediction, as well as providing experienced support and improvements to the T2K oscillation analysis. Responsibilities include taking positions as a Run Co-ordinator and DAQ expert for the T2K near detectors.

University of Oxford, Oxford, UK
PhD Student **October 2012–October 2016**
Research under the supervision of Prof. Alfons Weber, on the T2K experiment. Research focused on the use of a Markov Chain Monte Carlo method for oscillation analyses, in which multiple data samples, over a number of detectors, are fit simultaneously. Results include the first global analysis of neutrino and antineutrino data, producing the first ever 90% exclusion of CP conservation in the neutrino sector, and precise measurement of other neutrino oscillation parameters. Responsibilities to the collaboration include being a lead analyser for the oscillation analysis group and a DAQ expert for the T2K near detectors.

University of Oxford, Oxford, UK
Summer Student **July 2011–September 2011**
Oxford plasma theory group in conjunction with the Joint European Torus.

Paul Scherrer Institute, Switzerland
Summer Student **June 2010–September 2010**
Radiation safety and dosimetry group.

HONOURS AND AWARDS

- Winton Capital Prize for Statistical Analysis of Data, 2016
- Breakthrough Prize in Fundamental Physics, 2016
- Scholarship for Teaching and Research, Oriel College, University of Oxford, 2014, renewed 2015
- Pathfinders Award, Balliol College, University of Oxford, 2012
- Brink Scholarship, Balliol College, University of Oxford, 2011

COLLABORATION COMMITTEES AND RESPONSIBILITIES

- T2K paper committee for "Combined analysis of neutrino and antineutrino oscillations at T2K", published in Phys. Rev. Lett.
- T2K paper committee for "Measurement of Muon Antineutrino Oscillations with an Accelerator-Produced Off-Axis Beam", published in Phys. Rev. Lett.
- Run Co-ordinator for the T2K near detectors

• Data Acquisition Systems expert for the T2K near detectors

• Founder of T2K-Young, a T2K committee with the explicit aim of improving representation of young people within the collaboration

• Founder of T2K-Women, an informal group to enable networking of women in T2K and provide a forum to discuss gender and equality issues

CONFERENCE CONTRIBUTIONS

• **Current Status and Future Plans of T2K**
Presentation (plenary session); NuPhys 2016: Prospects in Neutrino Physics;
December 2016

• **Results and Prospects from T2K**
Presentation (plenary session); NuFact 2015: XVII International Workshop on Neutrino Factories and Future Neutrino Facilities; August 2015

• **Antineutrino Oscillation at the T2K Experiment**
Presentation (parallel session on neutrino oscillation); Institute of Physics Particle, Astroparticle, and Nuclear Physics Groups Conference 2015; March 2015

• **First Oscillation Analysis Using Neutrino and Antineutrino Data at T2K**
Poster (awarded Honourable Mention); Neutrino 2016: XXVII International Conference on Neutrino Physics and Astrophysics; July 2016

• **First Antineutrino Oscillation Results from T2K**
Poster; NuPhys 2015: Prospects in Neutrino Physics;
December 2015

SEMINARS AND COLLOQUIA

• **A Dummies' Guide to the T2K Oscillation Analyses**
T2K-Young Masterclass; May 2017

• **Neutrino Oscillation at T2K: A Bayesian Perspective**
Imperial College, London; High Energy Physics Seminar; November 2016

• **Search for Neutrino-Sector CP Violation in Neutrino and Antineutrino Oscillation at T2K**
Lancaster University; Particle Physics Seminar; November 2016

• **Antineutrino Oscillation at T2K**
Royal Holloway, University of London; Particle Physics Seminar; March 2015

TEACHING

• Oriel College tutor, University of Oxford: Electricity and Magnetism (2nd year undergraduate course), Subatomic Physics (3rd year undergraduate course),

Symmetry and Relativity (3rd year undergraduate course), 2014–2016

• Undergraduate admissions interviewer for Oriel College, University of Oxford, 2014-2015

• Undergraduate lab instructor: Nuclear Physics, 2013–2017

PUBLICATIONS

Publications in **bold** indicate a high level of personal contribution to the results reported through data analysis and paper writing.

PEER- REVIEWED PUBLICATIONS

• **K. Abe et al. (T2K Collaboration), Combined analysis of neutrino and antineutrino oscillations at T2K, *Phys. Rev. Lett.* 118, 151801, 2017**

• K. Abe et al. (T2K Collaboration), First measurement of the muon neutrino charged current single pion production cross section on water with the T2K near detector, *Phys. Rev. D* 96, 012010, 2017

• K. Abe et al. (T2K Collaboration), Measurement of coherent π^+ production in low energy neutrino-carbon scattering, *Phys. Rev. Lett.* 117, 192501, 2016

• K. Abe et al. (T2K Collaboration), Measurement of double-differential muon neutrino charged-current interactions on C8H8 without pions in the final state using the T2K off-axis beam, *Phys. Rev. D* 93, 112012, 2016

• **K. Abe et al. (T2K Collaboration), Measurement of muon antineutrino oscillations with an accelerator-produced off-axis beam, *Phys. Rev. Lett.* 116, 181801, 2016**

• **C. Wilkinson et. al., Testing CCQE and 2p2h models in the NEUT neutrino interaction generator with published datasets from the MiniBooNE and MINERvA experiments, *Phys. Rev. D* 93, 072010, 2016**

• K. Abe et al. (T2K Collaboration), Upper bound on neutrino mass based on T2K neutrino timing measurements, *Phys. Rev. D* 93, 012006, 2016

• K. Abe et al. (T2K Collaboration), Measurement of the muon neutrino inclusive charged- current cross section in the energy range of 1–3 GeV with the T2K INGRID detector, *Phys. Rev. D* 93, 072002, 2016

• K. Abe et al. (T2K Collaboration), Measurement of the ν_μ charged current quasielastic cross section on carbon with the T2K on-axis neutrino beam, *Phys. Rev. D* 91, 112002, 2015

• K. Abe et al. (T2K Collaboration), Measurement of

the electron neutrino charged-current interaction rate on water with the T2K ND280 π^0 detector, *Phys. Rev. D* 91, 112010, 2015
• K. Abe et al. (T2K Collaboration), Measurement of the ν_μ charged-current quasielastic cross section on carbon with the ND280 detector at T2K, *Phys. Rev. D* 92, 112003, 2015
• K. Abe et al. (T2K Collaboration), Measurement of neutrino oscillation in appearance and disappearance channels by the T2K experiment with 6.6e20 protons on target, *Phys. Rev. D* 91, 072010, 2015
• K. Abe et al. (T2K Collaboration), Neutrino oscillation physics potential of the T2K experiment, *Prog. Theor. Exp. Phys.* 043C01, 2015
• K. Abe et al. (T2K Collaboration), Search for short baseline ν_e disappearance with the T2K near detector, *Phys. Rev. D* 91, 051102(R), 2015
• K. Abe et al. (T2K Collaboration), Measurement of the inclusive ν_μ charged current cross section on iron and hydrocarbon in the T2K on-axis neutrino beam, *Phys. Rev. D* 90, 052010, 2014
• K. Abe et al. (T2K Collaboration), Measurement of the inclusive electron neutrino charged current cross section on carbon with the T2K near detector, *Phys. Rev. Lett.* 113, 241803, 2014
• K. Abe et al. (T2K Collaboration), Measurement of the neutrino-oxygen neutral-current interaction cross section by observing nuclear deexcitation gamma rays, *Phys. Rev. D* 90, 072012, 2014
• K. Abe et al. (T2K Collaboration), Measurement of the intrinsic electron neutrino component in the T2K neutrino beam with the ND280 detector, *Phys. Rev. D* 89, 092003, 2014; Erratum *Phys. Rev. D* 89, 099902, 2014
• K. Abe et al (T2K Collaboration), Precise measurement of the neutrino mixing parameter θ_{23} from muon neutrino disappearance in an off-axis beam, *Phys. Rev. Lett.* 112, 181801, 2014
• K. Abe et al. (T2K Collaboration), Observation of electron neutrino appearance in a muon neutrino beam, *Phys. Rev. Let.* 112, 061802, 2014
• K. Abe et al. (T2K Collaboration), Measurement of neutrino oscillation parameters from muon neutrino disappearance with an off-axis beam, *Phys. Rev. Lett.* 111, 211803, 2013

CONFERENCE PROCEEDINGS AND PRE- PRINTS

- **K. E. Duffy, Current status and future plans of T2K, Proceedings from Nu- Phys2016**, arxiv:1705.01764 [hep-ex], **2017**
- **K. E. Duffy, First oscillation analysis using neutrino and antineutrino data at T2K, Proceedings from Neutrino2016, submitted, publication forthcoming (available on request), 2016**
- **K. E. Duffy, First antineutrino oscillation results from T2K, Proceedings from NuPhys2015**, arxiv:1605.01626 [hep-ex], **2016**
- **K. E. Duffy, Results and prospects from T2K, in Proceedings of the XVII International Workshop on Neutrino Factories and Future Neutrino Facilities (NuFact15), Rio de Janeiro, 2015, edited by H. da Motta and J. Morfin, eConf C1508102, 11, 2016**
- K. Abe et al. (T2K Collaboration), Search for Lorentz and CPT violation using sidereal time dependence of neutrino flavour transitions over a short baseline, arxiv:1703.01361 [hep-ex], 2017
- K. Abe et al. (T2K Collaboration), Proposal for an extended run of T2K to 20×10^{21} POT, arxiv:1609.04111 [hep-ex], 2016
- K. Abe et al. (T2K Collaboration), Sensitivity of the T2K accelerator-based neutrino experiment with an Extended run to 20×10^{21} POT, arxiv:1607.08004 [hep-ex], 2016

TECHNICAL NOTES (AVAILABLE ON REQUEST)

- **K. E. Duffy, A. Kaboth, P. Dunne, L. Haegel, A Joint ND280-SK 1R$_\mu$-SK 1R$_e$ Fit of Neutrino and Antineutrino-mode Data using MCMC, *T2K Technical Note T2K-TN-269*,2016**
- **A. Kaboth, K. E. Duffy, R. Calland, D. Payne, Analysis of $\bar{\nu}_e$ Appearance using MCMC, *T2K Technical Note T2K-TN-247*, 2015**
- **A. Kaboth, K. E. Duffy, R. Calland, D. Payne, Analysis of Disappearance using MCMC, *T2K Technical Note T2K-TN-241*,2015**

Glossary

$1R_e$ Data sample at SK in which events are selected with only one reconstructed Cherenkov ring and particle identification consistent with an electron.

R_μ Data sample at SK in which events are selected with only one reconstructed Cherenkov ring and particle identification consistent with a muon.

2p–2h A neutrino interaction mode in which a neutrino scatters from a multi-nucleon state rather than a single nucleon. '2p–2h' stands for 'two particle–two hole', indicating that the neutrino interacts with a two-particle state, leaving two 'holes' in the nucleus in the final state. Term used interchangeably in this thesis with 'Meson Exchange Currents', or 'MEC'.

Antineutrino mode Beam configuration which produces a predominantly-$\bar{\nu}_\mu$ neutrino beam. Term used interchangeably in this thesis with 'Reverse Horn Current', or 'RHC'.

Asimov fake data set A fake data set in which the full MC prediction, without any statistical fluctuations, is used as the fake "data".

BANFF Beam and ND280 Flux extrapolation task Force, a group on T2K which fits only the ND280 data for those oscillation analyses which do not perform a simultaneous ND280-SK fit.

Bayes factor A statistic used for Bayesian hypothesis testing. The Bayes factor is equal to the likelihood ratio for two models, marginalised over all model parameters, and represents the weight of evidence from the data in favour of one model over another.

Bayes' theorem A theorem relating the joint probability distribution (i.e. the product of the likelihood and the prior probability distribution) and the posterior probability distribution. Given in Eq. 4.1.

CCQE Charged-Current Quasi-Elastic scattering. A neutrino interaction mode in which a neutrino scatters from an entire nucleon, rather than its constituent particles, in a 2-particle→2-particle interaction. This is the dominant neutrino interaction process at T2K, and the interaction channel which is used for the oscillation analysis.

Confidence interval Frequentist method of defining the uncertainty on a point estimate. A 90% confidence interval is defined such that, should the experiment be repeated 100 times, 90% of the best-fit points would lie within the interval.

K.E. Duffy, *First Measurement of Neutrino and Antineutrino Oscillation at T2K*,
Springer Theses, DOI 10.1007/978-3-319-65040-1

Credible interval Bayesian method of defining the uncertainty on a point estimate. A 90% credible interval is defined such that it contains 90% of the posterior probability from a single experiment.

DIS Deep Inelastic Scattering. A neutrino interaction mode in which a neutrino interacts with the constituent particles of a nucleon, rather than the nucleon itself. When used in this thesis, refers to charged-current deep inelastic scattering.

ECal Electromagentic Calorimeter, subdetector of ND280. ND280 contains two ECals: the P0D ECal (which surrounds the P0D) and the tracker ECal (which surrounds the TPCs and FGDs).

E_{rec} Reconstructed neutrino energy, calculated under the assumption that a given interaction was CCQE scattering of a neutrino or antineutrino from a stationary bound nucleon (using a binding energy of 27 MeV for interactions on Oxygen at the far detector). The formula used to calculate the reconstructed neutrino energy is given in Eq. 3.1.

FGD Fine-Grained Detector, subdetector of ND280. ND280 contains two FGDs, which provide a target for neutrino interactions: FGD1 (target material: hydrocarbon) and FGD2 (target materials: hydrocarbon and water).

FHC Forward Horn Current, a beam configuration which produces a predominantly-ν_μ neutrino beaem. Term used interchangeably in this thesis with 'neutrino mode'.

FSI Final State Interaction, in which a particle produced in a neutrino interaction re-interacts in the nucleus before it is detected.

HPD credible interval Highest Posterior Density credible interval. A method of defining credible intervals which requires that every point inside the credible interval has a larger posterior probability density than every point outside the interval. This definition is used to construct the credible intervals presented in this thesis.

IH Inverted mass hierarchy, the neutrino mass ordering in which $m_3^2 < m_2^2$.

INGRID On-axis near detector for T2K. Located 280 m from the neutrino production point, and on-axis with respect to the neutrino beam.

Joint probability distribution, $P(D, \vec{\theta})$ Probability distribution for both the data D and model parameters $\vec{\theta}$. Equal to a product of the prior distribution for the model parameters, $P(\vec{\theta})$, and a likelihood, $P(D|\vec{\theta})$.

LFG Local Fermi Gas, a nuclear model used to simulate neutrino-nucleus interactions.

Likelihood, $P(D|\vec{\theta})$ Probability of observing some set of data D given some model parameter values $\vec{\theta}$.

Marginalisation Method for eliminating nuisance parameters from the posterior distribution by integrating the posterior distribution over all unwanted parameters.

MC Monte Carlo simulation.

MCMC Markov Chain Monte Carlo, the fitting method used in this thesis.

MEC Meson Exchange Currents. A neutrino interaction mode in which a neutrino scatters from a multi-nucleon state rather than a single nucleon. Term used interchangeably in this thesis with '2p–2h'.

Micromegas detector A gaseous detector structure which has high granularity and is capable of handling high signal rates [1,2].

MPPC Multi-Pixel Photon Counter. Used to detect scintillation light collected on WLS fibres in INGRID and ND280 [3].

NA61/SHINE A dedicated hadron-production experiment at the CERN SPS which studies the particles produced when a 30 GeV proton beam interacts with a graphite target. Data measurements from NA61/SHINE are used to tune the T2K flux prediction.

ND Near Detector. Usually used in this thesis to refer to the T2K off-axis near detector, ND280, which is used directly in the oscillation analysis fit.

ND280 Off-axis near detector for T2K. Located 280 m from the neutrino production point, at the same off-axis angle with respect to the neutrino beam as the far detector (2.5°).

NEUT Software used to simulate neutrino interactions and produce MC predictions at the near and far detectors.

Neutrino mode Beam configuration which produces a predominantly-ν_μ neutrino beam. Term used interchangeably in this thesis with 'forward horn current', or 'FHC'.

NH Normal mass hierarchy, the neutrino mass ordering in which $m_3^2 > m_2^2$.

NIWG Neutrino Interaction Working Group, a group on T2K which studies neutrino interaction models using T2K and external data, and provides recommendations for cross-section parameterisations and uncertainties used in the oscillation analysis.

Nuisance parameter A model parameter that affects the probability distribution but which we don't care about for its own sake (for example, a systematic parameter). The posterior distribution is often marginalised over nuisance parameters to find the probability only as a function of the parameters of interest.

P-Theta One of the two other oscillation analysis groups on T2K, against whom this analysis has been checked as part of the validation process.

P0D π^0 detector, subdetector of ND280.

PID Particle Identification.

PMT Photo-Multiplier Tube. Used to detect Cherenkov light produced in SK.

Posterior distribution, $P(\vec{\theta}|D)$ Probability distribution for some model parameters $\vec{\theta}$ given observed data D.

POT Protons On Target, a measure of the amount of data collected by T2K.

Prior probability distribution, $P(\vec{\theta})$ Probability distribution for the model parameters, $\vec{\theta}$, which contains all prior knowledge about the parameters in the model before the data are analysed.

RES Resonance production, a neutrino interaction mode which often produces a single pion in the final state. When used in this thesis, refers to charged-current resonance production.

RFG Relativistic Fermi Gas, a nuclear model used to simulate neutrino-nucleus interactions.

RHC Reverse Horn Current, a beam configuration which produces a predominantly-$\bar{\nu}_\mu$ neutrino beam. Term used interchangeably in this thesis with 'antineutrino mode'.

RPA Random Phase Approximation, a nuclear model used to simulate neutrino-nucleus interactions. Two RPA models are available: relativistic RPA and non-relativistic RPA.

SF Spectral Function, a nuclear model used to simulate neutrino-nucleus interactions.

SI Secondary Interaction, in which a particle produced in a neutrino interaction re-interacts in the detector before it is measured.

SK/Super-K Super-Kamiokande, the far detector for T2K. Located 295 Km from the neutrino production point, off-axis by 2.5° with respect to the neutrino beam.

SMRD Side Muon Ranging Detector, subdetector of ND280.

T2K-SK The group on T2K with responsibility for the SK data, MC prediction, and detector systematic uncertainties.

TPC Time Projection Chamber, subdetector of ND280. ND280 contains three TPCs, which are used for particle identification and to measure the momentum of products of neutrino interactions in the FGDs.

Tracker region of ND280 Collective name for the 2 FGDs and 3 TPCs in ND280.

Trip-T front-end electronics boards Electronics boards used to read out signals in INGRID and some of the ND280 subdetectors. Based on the Trip-T chip, which was originally designed at Fermilab for the Tevatron D0 experiment [3,4].

VaLOR One of the two other oscillation analysis groups on T2K, against whom this analysis has been checked as part of the validation process.

WLS fibre Wavelength-Shifting fibre. Used to collect light produced in the scintillator material of INGRID and ND280.

References

1. Y. Giomataris, Ph. Rebourgeard, J.P. Robert, G. Charpak, MICROMEGAS: a high-granularity position-sensitive gaseous detector for high particle-flux environments. Nucl. Instr. Methods Phys. Res. Sect. A Accel. Spectro. Detec. Assoc. Equ. **376**(1), 29–35 (1996). 10.1016/0168-9002(96)00175-1
2. I. Giomataris et al., Micromegas in a bulk. Nucl. Instr. Methods Phys. Res. Sect. A Accel. Spectro. Detec. Assoc. Equ. **560**(2), 405–408 (2006). 10.1016/j.nima.2005.12.222
3. A. Vacheret et al., The front end readout system for the T2K-ND280 detectors, in *2007 IEEE Nuclear Science Symposium Conference Record*, vol. 3, pp. 1984–1991 (2007). 10.1109/NSSMIC.2007.4436543
4. J. Estrada, C. Garcia, B. Hoeneisen, P. Rubinov, MCM II and the trip chip, Technical Report 2226, FERMILAB (2002), https://inspirehep.net/record/635982/files/fermilab-tm-2226.PDF

MIX
Papier aus verantwortungsvollen Quellen
Paper from responsible sources
FSC® C105338

If you have any concerns about our products,
you can contact us on
ProductSafety@springernature.com

In case Publisher is established outside the EU,
the EU authorized representative is:
Springer Nature Customer Service Center GmbH
Europaplatz 3, 69115 Heidelberg, Germany

Printed by Libri Plureos GmbH
in Hamburg, Germany